Getting Started with the Photon

Simon Monk

MAKER MEDIA™
SAN FRANCISCO, CA

Make: Getting Started with the Photon

by Simon Monk

Copyright © 2015 Maker Media. All rights reserved.

Printed in the United States of America.

Published by Maker Media, Inc., 1160 Battery Street East, Suite 125, San Francisco, CA 94111.

Maker Media books may be purchased for educational, business, or sales promotional use. Online editions are also available for most titles (*http://safaribooksonline.com*). For more information, contact our corporate/institutional sales department: 800-998-9938 or *corporate@oreilly.com*.

Editor: Roger Stewart
Production Editor: Kristen Brown
Proofreader: Sharon Wilkey
Interior Designer: David Futato
Cover Designer: Karen Montgomery
Illustrator: Rebecca Demarest
Technical Reviewers: Dr. Stephen Hall and Brett Walach

May 2015:   First Edition

Revision History for the First Edition

2015-04-24:   First Release

See *http://oreilly.com/catalog/errata.csp?isbn=9781457187018* for release details.

978-1-457-18701-8

[LSI]

Contents

Foreword

According to *MIT Technology Review*, 2013 was the year of the Internet of Things. The following year, Cisco said 2014 was the year of the Internet of Things. This year, CNBC says that 2015 is the year of the Internet of Things. It is likely that every year for the next decade will be the "year of the Internet of Things." But what does that mean, exactly?

The Internet of Things (IoT) is a broad concept that suggests that as the cost of connectivity comes down, more and more objects around us will be Internet-connected. We're used to thinking about the Internet in terms of devices with screens: your personal computer, for instance, and your smartphone. The Internet of Things covers devices that you wouldn't typically think about being connected to the Internet, and includes categories like wearables (Fitbit, smart watches), smart home (connected lights, appliances, and toys), the industrial Internet (feedback systems on wind turbines), smart cities (connected parking meters and traffic lights), smart farms (connected irrigation systems), and more. It's a category that is difficult to define because of its breadth, but the common thread is that things that weren't connected before are becoming connected now.

I believe that the Internet of Things is the third wave of personal computing (the first being the PC, and the second being smartphones). And, unlike the tech press, I don't think we've reached the "year of the Internet of Things" quite yet. I believe we're still in the prototyping days, where even most IoT products coming to market are just the first experiments that will give way to a large, vibrant ecosystem in a few years' time. This is like the early 90s for the Internet, or the early 2000s for smartphones.

And, like PCs and smartphones, the rapid growth that we're seeing now—and that we'll continue to see for the next decade—will present massive opportunities for the people who jump in early.

Industries will change, jobs will be created, and fortunes will be made and lost.

This book is about a tool: the Photon. The Photon is a simple development kit, but it represents a starting point that can take you much further. Your journey may start with the Photon and end with a connected garage-door opener for your house. Or it may end with a connected garage-door opener that you bring to market and sell to hundreds of thousands of people around the world.

For you, dear reader, the "year of the Internet of Things" will be the year that you join in. Every great company starts as an idea, and every great product starts as a prototype. What will you build first?

—Zach Supalla, Founder and
CEO of Particle, April 2015

Preface

What's in a Name?

This book has been written at a time when the Spark organization was rebranding itself as Particle. As a result of this, you will find that some of the screenshots have the old company name, as they were taken before the new web interface became available. You will also find that the code for this book uses class names and third-party code libraries that still show the Spark name. These are set to work in the short term, but when they eventually become deprecated, watch for information on these changes at the Particle website (*http://www.particle.io*).

Conventions Used in This Book

The following typographical conventions are used in this book:

Italic

> Indicates new terms, URLs, email addresses, filenames, and file extensions.

`Constant width`

> Used for program listings, as well as within paragraphs to refer to program elements such as variable or function names, databases, data types, environment variables, statements, and keywords.

`Constant width bold`

> Shows commands or other text that should be typed literally by the user.

`Constant width italic`

> Shows text that should be replaced with user-supplied values or by values determined by context.

This element signifies a general note, tip, or suggestion.

 This element indicates a warning or caution.

Using Code Examples

The code examples are all available as a code library directly from the Web IDE. They are also available for download at *https://github.com/simonmonk/photon_book*.

This book is here to help you get your job done. In general, you may use the code in this book in your programs and documentation. You do not need to contact us for permission unless you're reproducing a significant portion of the code. For example, writing a program that uses several chunks of code from this book does not require permission. Selling or distributing a CD-ROM of examples from Make: books does require permission. Answering a question by citing this book and quoting example code does not require permission. Incorporating a significant amount of example code from this book into your product's documentation does require permission.

If you feel your use of code examples falls outside fair use or the permission given here, feel free to contact us at *bookpermissions@makermedia.com*.

We appreciate, but do not require, attribution. An attribution usually includes the title, author, publisher, and ISBN. For example: "*Make: Getting Started with the Photon* by Simon Monk (Maker Media). Copyright 2015 Maker Media, 978-1-4571-8701-8."

Safari® Books Online

Safari Books Online is an on-demand digital library that delivers expert content in both book and video form from the world's leading authors in technology and business.

Technology professionals, software developers, web designers, and business and creative professionals use Safari Books Online as their primary resource for research, problem solving, learning, and certification training.

Safari Books Online offers a range of plans and pricing for enterprise, government, education, and individuals.

Members have access to thousands of books, training videos, and prepublication manuscripts in one fully searchable database from publishers like Maker Media, O'Reilly Media, Prentice Hall Professional, Addison-Wesley Professional, Microsoft Press, Sams, Que, Peachpit Press, Focal Press, Cisco Press, John Wiley & Sons, Syngress, Morgan Kaufmann, IBM Redbooks, Packt, Adobe Press, FT Press, Apress, Manning, New Riders, McGraw-Hill, Jones & Bartlett, Course Technology, and hundreds more. For more information about Safari Books Online, please visit us online.

How to Contact Us

Please address comments and questions concerning this book to the publisher:

> Make:
> 1160 Battery Street East, Suite 125
> San Francisco, CA 94111
> 877-306-6253 (in the United States or Canada)
> 707-639-1355 (international or local)

Make: unites, inspires, informs, and entertains a growing community of resourceful people who undertake amazing projects in their backyards, basements, and garages. Make: celebrates your right to tweak, hack, and bend any technology to your will. The Make: audience continues to be a growing culture and community that believes in bettering ourselves, our environment, our educational system—our entire world. This is much more than an audience; it's a worldwide movement that Make: is leading—we call it the Maker Movement.

For more information about Make:, visit us online:

Make: magazine: *http://makezine.com/magazine*
Maker Faire: *http://makerfaire.com*
Makezine.com: *http://makezine.com*
Maker Shed: *http://makershed.com*

We have a web page for this book, where we list errata, examples, and any additional information. You can access this page at *http://bit.ly/make_gs_photon*.

To comment or ask technical questions about this book, send email to *bookquestions@oreilly.com*.

1/The Photon

In this chapter, you will learn a little about the Internet of Things in general, as well as the Photon in particular. The Photon and its older brother, the Spark Core, are explored along with some background about where it has come from and where it sits in the pantheon of developer boards.

The Internet of Things

It used to be that the only way you could interact with the Internet was to use a web browser on your computer. A browser would allow the computer to send requests to a web server that would send back information to be displayed.

The browser would display this information using a computer monitor, and the user would type text on her keyboard and follow hyperlinks with the click of a mouse. As far as inputs and outputs were concerned, those were your options.

The Internet of Things (IoT) has changed all this. Now all sorts of sensors and appliances can be connected to the Internet. The IoT encompasses a wide range of systems:

- Home automation systems that control lighting, heating, and doors by using web browser or network-enabled smartphone applications. These may be used to control systems over the local area network, or over the Internet using WiFi or a cellular network.
- Arrays of sensors, such as the Safecast open radiation-monitoring system that was developed following the Fukushima nuclear disaster.

Products and maker projects that will become part of the IoT are springing up all over the place. These include successful projects like the Nest smart thermostat as well as many IoT products that use the accelerometer, location services, and communication features of smartphones to capture information about people's health and activity levels.

Since so many people are creating IoT projects, it makes perfect sense to provide a simple modular framework for both hardware and software that provides an easy-to-use IoT technology kit. This is exactly where the Particle team comes in. They provide IoT technology in a box—a very small and low-cost box. What's more, the technology is easy to use, open source, and based on the very popular Arduino software framework.

Sparks in the Clouds

The hardware component of this IoT framework is the Photon. The Photon is the next generation of Particle's IoT platform that began with the Spark Core. The Photon is backward compatible with the Spark Core, and so most of what is detailed in this book about the Photon will also work with the older Core.

Although other technologies exist to help you build IoT devices, they often neglect the all-important software framework that allows the device to communicate with other devices and browsers over the Internet. The Particle approach, by contrast, integrates the hardware and software seamlessly.

Figure 1-1 shows how a typical IoT device built using a Photon or Core might interact with the Internet.

An IoT device using a Photon/Core might provide remote unlocking of a door. In such a case, a user would access a web page on a browser that has an Unlock button. This page will have been served from a web server somewhere on the Internet. When the user clicks the Unlock button in the browser, the browser sends a message to the cloud service that forwards the message along to the Photon running inside the connected device. The Photon/Core controlling the electromechanical door latch then knows it should unlock the door.

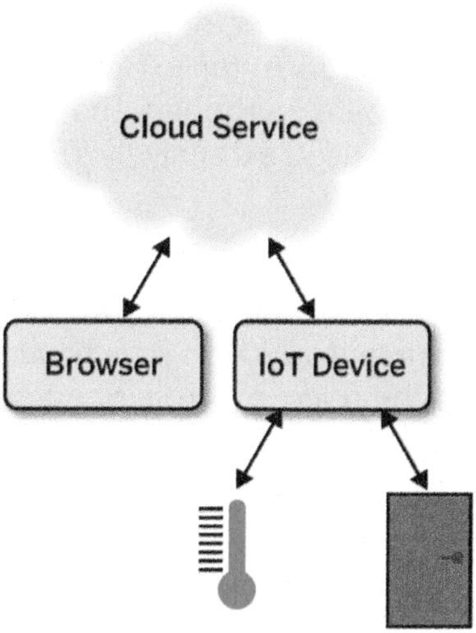

Figure 1-1. *Internet of Things communication*

If, on the other hand, the IoT device were acting as a sensor—let's say, for temperature—then the Photon/Core could send temperature readings to a cloud service. Those readings could be stored temporarily until the user's browser has a chance to pick them up and display the latest reading on the browser window.

To use Particle's cloud service, you first register online with Particle and then identify each of your Photons/Cores, which will have registered themselves with the cloud service as being yours. All the Photon needs to do to register itself is to have access to your WiFi network. This process not only allows you to ensure that you know which Photon or Core you are interacting with but also allows you to program your Photons and Cores over the air from the comfort of your web browser.

Other IoT Platforms

Before plunging into the delightfully warm and pleasant waters of the Photon pool, it's worth exploring some of the Photon's competitors. This will also reveal something of the motivations behind the design of the Photon.

The Photon is of course not the only IoT device around. In fact, the single most used platform for IoT development is the Arduino microcontroller board, although the Raspberry Pi single-board computer is also used extensively in IoT projects.

Arduino

Microcontrollers are essentially low-powered computers on a chip. They have input/output (I/O) pins to which you can attach electronics so that the microcontroller can, well, *control* things. The Arduino is a simple-to-use and low-cost, ready-made board that allows you to make use of a microcontroller in your projects.

The Arduino has become the platform of choice for makers and hackers looking for a microcontroller to use, and the most common Arduino model is the Arduino Uno.

The popularity of Arduino is due to many factors:

- Low cost (around $25 for an Arduino Uno)
- Open source hardware design—there are no secrets to its design and built-in software
- Easy-to-use integrated development environment with which to program the Arduino
- Plug-in shields that plug onto the top of the Arduino and add features such as displays and motor drivers

There is, of course, one factor that makes an Arduino Uno by itself useless as an IoT device: it has no network connection, either wired or wireless. You either need to use one of the specialized Arduino models that include an Ethernet network port (such as the Arduino Ethernet or Yun) or add a WiFi or Ethernet shield to the Arduino that then gives it the network connection it needs to communicate over the Internet. This adds considerably to the size and cost of your project.

Figure 1-2 shows an Arduino Uno with a WiFi shield attached. The total cost of this combination is over $100.

Figure 1-2. *An Arduino Uno with WiFi shield*

Another possibility is to use the Arduino Yun. This device is the same size as an Arduino Uno but has a built-in WiFi module. On the face of it, this provides similar hardware capabilities to the Photon, but at a much higher price of around $75.

All of these Arduino-based solutions suffer from one major disadvantage as an IoT platform, and that is software. They provide the base capabilities to communicate with the Internet but do not offer any software framework to make it easy to create IoT projects without a lot of tricky network programming.

Later, you will see how the Photon borrows many of the concepts of Arduino, including its programming language, but then provides a software framework with which to build your IoT projects, all at a much lower cost than Arduino can compete with.

Raspberry Pi and BeagleBone

The Raspberry Pi and BeagleBone Black (Figure 1-3) are both single-board computers, about the size of a credit card, that run the Linux operating system. They have USB ports and HDMI video output, so you can set them up with a keyboard, mouse, and monitor and use them just like a regular computer.

Figure 1-3. *A Raspberry Pi and BeagleBone Black*

The Raspberry Pi is shown on the left of Figure 1-3 and the BeagleBone Black on the right. Both boards can use low-cost USB WiFi adaptors and have I/O pins to control electronics and interface with sensors, making them quite suitable for IoT projects.

Although both boards are low in cost, the Raspberry Pi from $35 and the BeagleBone from $55, they are quite large (compared to a Photon) and generally contain a lot more than you need for a simple IoT project.

Intel Edison

Intel has developed a small Linux-based board called the Edison. The Edison is designed to be embedded into IoT projects and is perhaps the most direct competition to the Photon.

The Edison is small but is considerably more expensive than the Photon. It has a delicate 70-contact connector that requires a separate breakout board if you want to start attaching external electronics to it. There are several such prototyping boards available, the most popular of which is an Arduino-compatible board.

Although receiving a lot of interest in the Maker community, this is probably a board that will lend itself best to high-end or professional use, not least because the device is a lot harder to get started with.

A Tour of the Photon

Figure 1-4 shows a Photon with some of its main features labelled.

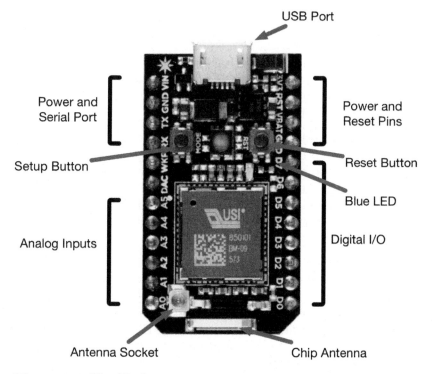

Figure 1-4. *The Photon*

The two buttons on the Photon, Setup and Reset, let you set new WiFi credentials and restart your device, and they can be used together to trigger a full factory reset.

At the top of the board, you will find the micro-USB port. Although the main point of this is to provide power to the Photon, it can also be used for USB programming of the board (see Chapter 9) and for USB communication with a computer.

Moving around the board clockwise, you'll find the Power and Reset pins (including 3v3, RST, VBAT, and GND). The Photon converts input power provided via the micro-USB power or the VIN pin into a 3.3-volt supply used by the board; this is available via the topmost pin, 3v3. The logic operates at 3.3V rather than the 5V that you might be used to as an Arduino user. The RST pin can be used much like the Reset button, when pulled low (connected to Ground). This can be useful if your project will be in an enclosure, but you are fairly unlikely ever to want to use this pin.

The VBAT pin allows a small backup battery (or supercapacitor) to be attached to the Photon, to power it while it is in deep sleep mode to preserve its memory, so that when it wakes, it can continue from when it went to sleep.

The pins labelled D0 to D7 are general-purpose input/output (GPIO) pins that can act as either digital inputs or outputs (see Chapter 4). Some of these pins can also act as analog outputs (pins D0 to D3) using a technique called pulse-width modulation (PWM). There is a blue LED next to pin D7 that is connected to D7. You can turn this on and off from your programs, or using the Tinker app on your smartphone.

The Photon has a built-in chip antenna that will work fine in most WiFi situations, but the Photon also has a tiny antenna socket to which an external antenna can be attached. This is useful for extending the WiFi range of the device by adding a more sensitive or directional antenna. By default the Photon will attempt to choose the best antenna, but you can also control which antenna is preferred in your firmware.

The pins A0 to A5 are primarily analog inputs that can measure a voltage between 0 and 3.3V. These will typically be used for sensors. For example, in Chapter 6 you will use a light sensor with one of these pins. The analog pins can also be used as digital inputs or outputs, just like the pins D0 to D7 and, like the digital pins, some of the analog pins (A4, A5) can also be used as PWM analog outputs.

The DAC (Digital to Analog Converter) pin is a special analog output pin. This is a true analog output that can be set to any voltage between 0 and 3.3V.

The WKP (Wakeup) pin is used to wake up the Photon after it has put itself into a deep sleep mode. Writing software that puts the Photon to sleep some of the time makes it practical to run the Photon from batteries where power consumption needs to be minimized. WKP, RX and TX can also be used as GPIO.

The TX and RX (Transmit and Receive) pins are used for serial communication. This is useful for connecting a Photon to certain types of peripherals, such as GPS modules, that have a serial interface.

Above these pins is a second GND pin and the VIN pin. You can supply between 3.6V and 5.5V to the VIN pin to power the board as an alternative to using the USB port.

The Spark Core vs. Photon

The Photon is not the first IoT board to be produced by Particle. That honor belongs to the Spark Core. Both boards are shown side by side in Figure 1-5, and as you can see, the boards look very similar to each other.

The main differences are that the Photon uses a different WiFi chip (Broadcom rather than Texas Instruments). It also uses a faster processor with more RAM.

The pins are mostly the same between the two boards; however, there are a few important changes to be aware of. First, the Core does not have DAC and WKP pins; instead it has two analog inputs, A6 and A7.

Figure 1-5. *The Spark Core (left) and Photon (right)*

The examples in this book have been tested on both the Photon and the Core, so if you have one of the older Cores, you should be just fine.

Programming

Particle has gone to great lengths to make programming the Photon and Core as easy as possible, both by providing a very simple-to-use development environment and by choosing the Arduino C programming language as the basis for the Photon's programming language.

Since the Photon is an Internet device, it makes perfect sense to program the device over the Internet, so most of the time, you will write the code for your Photon in your web browser and then push it out to your Photon, which will be looking out for updates. An offline programming environment is also available for more advanced users (see Chapter 9).

For those thinking of using the Photon for commercial applications, the ability to update over the air is incredibly powerful, as it allows remote updates of the Photon's software from anywhere in the world.

This may start a few alarm bells ringing. After all, if the Photon were controlling your home's front door or heating, you would not want someone updating your Photons and Cores, telling them to unlock your doors and turn the heating up to maximum. Fortunately, several security mechanisms are used to secure the Photon. These include the use of Secure Sockets Layer (SSL) communication as well as authentication keys that ensure that only you can update your Cores.

Figure 1-6 shows the Web IDE (integrated development environment) for the Photon/Core.

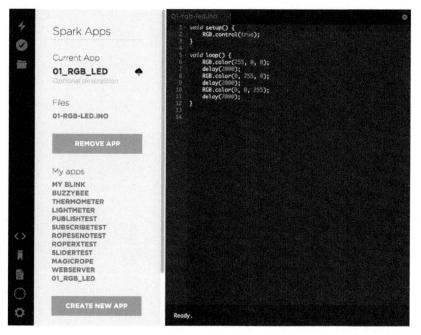

Figure 1-6. *The Particle Web IDE*

The Web IDE allows you to write code that you can then send down to one of your Photons over the Internet.

Summary

By now, you must be itching to get started with the Photon, so in Chapter 2, you will learn how to get up and running with your Photon and install your first program onto it.

2/Quick Start

In this chapter, you will launch right into using your Photon or Core. If you are new to programming and this kind of technology, then just work through the instructions here and do not worry too much if there are things that you don't understand; these will all be explained in much more detail later in the book.

Signing Up

One of the things that you get when you buy a Particle device is access to Particle's cloud service. You will need to register to gain access to this. Registering with the cloud service is essential, as it allows you to manage your Photons and Cores over the Internet, including programming them remotely.

The cloud service is also what a Photon/Core uses to communicate with other Photons, Cores, and browser applications.

Your first step should therefore be to sign up at *http://www.particle.io*. Click the BUILD button to begin the sign-up process. Don't worry, this is strictly just to use your Photon over the Internet; you will not be bombarded with spam.

Once you are all signed up, you will find yourself in the Web IDE, ready to start writing programs. Before you can do that, you need to connect your Photon to the Internet.

Connecting to WiFi

Although it is possible to program a Photon or Core over its USB connection (see Chapter 9), it makes more sense to program an Internet device like this over the Internet by using the Web IDE. To do that, the Photon or Core needs to be connected to the Internet over WiFi. To be able to connect using WiFi, it needs to know the SSID (network name) and passphrase of your home network.

As you may have noticed, the Core and Photon have no keyboard or display, so you need a way of telling them these details.

WiFi Networks with Login Screens

Connecting to WiFi in a home environment with direct access to the Internet is no problem for the Photon or Core. Even a WiFi hotspot setup using your cellphone should work.

If, however, you are trying to use a Photon or Core at, say, a hotel or public network requiring you to fill in a form or use a web-based login screen after joining the WiFi network, then you probably won't be able to connect.

Connecting a Core

If you have a Photon, rather than a Core, you need to skip ahead to the next section.

To help you get your Core connected, Particle has produced a smartphone app available for Android and iPhone. The Particle mobile app, also known as the Tinker app in the documentation, goes by the name Spark Core in the App Store and Play Store. The app does more than allow you to connect your Core to the Internet; it also allows you to control the Core, turn pins on and off, etc. If you do not have a suitable smartphone, all is not lost; you can configure both the Core and Photon by using the USB cable (see Chapter 9).

When you plug in a brand-new Core, it should blink dark blue, indicating that it is "listening" for a WiFi network to connect to. Unless it is in this state, you will not be able to connect it to your wireless network. You can put your device back into this state either by clearing credentials (hold Setup until the Core blinks blue very rapidly), or by doing a factory reset on it as described in Chapter 9.

You will need an Android or iOS (iPhone/iPad) device to use the Tinker app to connect your device, and the phone or tablet must be connected to the WiFi network that you want to join the Core to.

When you first start the Tinker app, you will be prompted to log into your Particle account (Figure 2-1). Sign in using the same email and password that you gave when you signed up to Particle.

Figure 2-1. *Logging into your Particle account*

After you are logged in, you will be prompted to enter your WiFi password. The SSID (wireless network name) will be prefilled, as your device is already connected to that wireless network (Figure 2-2).

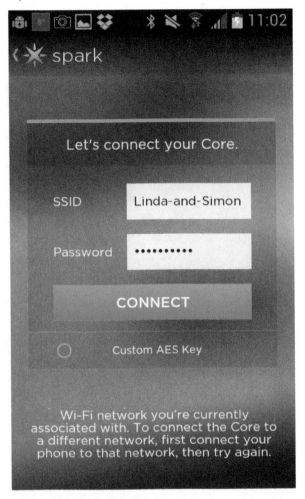

Figure 2-2. *Entering your WiFi password*

The app should then discover your Core and prompt you to supply a name for your Core (Figure 2-3). At this point, the Core will be "shouting rainbows at you," as it says in the dialog box.

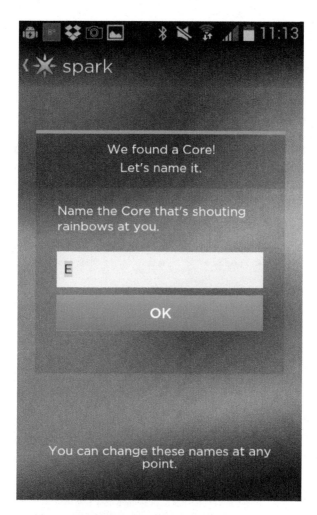

Figure 2-3. *Naming your device*

The Core should then reset, and when it's finished, the LED should be gently "breathing" in cyan and ready to use.

Connecting a Photon

Connecting a Photon works in much the same way as connecting a Core, but behind the scenes the Photon will be briefly hosting its own WiFi network. The mobile app connects to this network and securely provides the credentials to the device. The

Photon needs a different mobile app (the Photon app) to set up your Photon.

The steps that will take place in order to get the network name and password onto a Photon are as follows:

1. Connect your phone to the Photon's temporary WiFi network (soft access point).
2. Select the network whose details you want to pass to the Photon.
3. Let the Photon connect to your network and then register itself on the Particle web service.

After logging in, the app will remind you to power up your Photon (Figure 2-4).

Figure 2-4. *Prepare your Photon*

The next step involves temporarily connecting your phone to the soft access point being hosted by the Photon. This means you will need to use your iOS or Android device's settings to change networks. The Photon app will prompt you with the name of the network that you should connect to (Figure 2-5).

Figure 2-5. *Finding the temporary network name*

Figure 2-6 shows the settings screen for an iPhone with the Photon's soft access point (Photon-F1FC) selected.

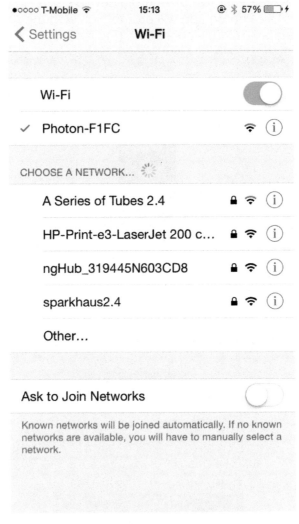

Figure 2-6. *Connecting your phone to your Photon's soft access point*

When you have done this, you will be returned to the Photon app, where you now have to select your home WiFi network from the list (Figure 2-7).

Figure 2-7. *Selecting your network for the Photon to connect to*

You will then be asked to enter the credentials so that the Photon can join the network (Figure 2-8).

Figure 2-8. *Entering credentials*

Finally, the Photon will connect itself to the Particle cloud service (Figure 2-9).

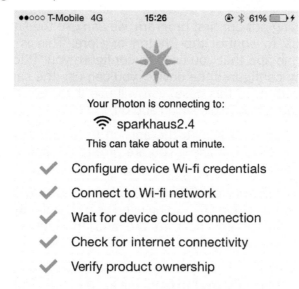

Figure 2-9. *A Photon connecting and registering*

Controlling Pins with the Tinker App

Before trying out our first program, we can use the smartphone Tinker app to control the Photon or Core. This is the same smartphone app that you used to configure your Photon/Core, but having configured the device, you can use the same app to interact with it. In this case, you will use it to turn the built-in blue LED attached to pin D7 on and off.

The first step is to launch the app on your Android phone or iPhone. Log in to Particle if you are prompted, and you should find yourself at the main control screen.

Before you can turn pin D7 on and off, you first have to tell the Photon/Core that pin D7 is to be used as a digital output; so tap and hold the screen next to D7, and the menu shown in Figure 2-10 should appear.

Select the option digitalWrite. Now, if you tap the D7 pin on the screen, the blue LED next to pin 7 will toggle on and off.

Project 1. Blink the Tiny Blue LED

Programmers have a tradition that the first program that they write in any language should be called Hello World. This usually involves just getting the program to display the words *Hello World*. Clearly the Photon/Core doesn't have anywhere to display those words, so we will have to think of something else as our easiest possible program.

Arduino and other similar boards have adopted the task of blinking an LED as the most basic type of thing that you might want to do with a board like the Photon, so let's start with that.

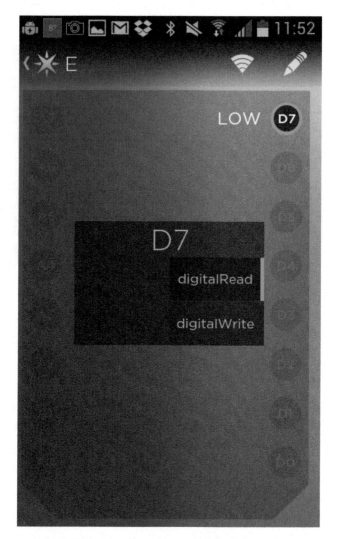

Figure 2-10. *Controlling D7 with the Tinker app*

Log in to particle.io by clicking the LAUNCH button, and you should find yourself at the Web IDE page. At the bottom left of the screen, you will find a section called Example Apps. Click the option BLINK AN LED, and the program shown in Figure 2-11 will load into the IDE's editor.

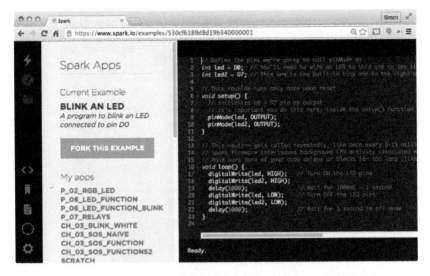

Figure 2-11. *The Web IDE with the blink program loaded*

Do not worry about how this program works; all will be explained in the next chapter. For now we just want to see that LED blink, so click the icon in the top left of the IDE that looks like a lightning bolt. This is the Flash button, which will flash the blink program onto your Photon.

You should see a message appear in the status area at the bottom of the screen that says: *Flash successful! Please wait a moment while your Photon is updated...*

Occasionally, the flashing will appear to hang, and nothing happens. Eventually, the IDE will time out, and you can try again. It may help to reset the Photon/Core by clicking the Photon's Reset button.

At this point, the RGB LED on the device should start blinking away in purple, indicating that the Photon/Core is being updated. After the device has received the program, it will reset itself (flashing green), and the RGB LED will blink green while it reconnects to your WiFi. It should resume breathing cyan when it connects to the cloud. Once the Photon/Core restarts, the tiny blue LED next to pin 7 will flash on and off every two seconds.

Occasionally, connection problems occur, and the flashing process fails. If this happens, just try again.

You will meet this program in the next chapter, where all will be explained; but before that, in Project 2, you will learn how to create a new program and type in the code for it.

Project 2. Control the Photon's LED

Just as you can control the tiny blue LED attached to pin D7 of the Photon or Core, you can also control the RGB LED that the Photon uses to indicate its status. In this mini-project, you will type a short program into the Web IDE that will make the RGB LED light up red, then change to green, and then, after a further two seconds, turn blue, and so on.

The first step is to click the CREATE NEW APP button on the lefthand side of the Web IDE. This will start a new app, as shown in Figure 2-12.

Figure 2-12. *Creating a new app*

As you can see, you are being prompted for a name for the app. This needs to be unique, and whatever you type here as a name will be converted into uppercase letters. If you want to separate the words of the app name, you can use the underscore (_) character. Let's call the app RGB_LED.

When you create a new project, the Web IDE will automatically create a blank template into which you can write your code. At the moment, the program for the app looks like this:

```
void setup() {
}
void loop() {
}
```

Change this text so that it appears as follows:

```
void setup() {
    RGB.control(true);
}

void loop() {
    RGB.color(255, 0, 0);
    delay(2000);
    RGB.color(0, 255, 0);
    delay(2000);
    RGB.color(0, 0, 255);
    delay(2000);
}
```

All the text that you are going to add into the template needs to be tabbed in one tab space, so on each new line, press the Tab key once before typing the line of code. Notice that each line of code must end in a semicolon.

When you have typed in all of the code, save the app by clicking the Save icon (which looks like a folder), and then hit the Flash button (lightning bolt) to program the Photon.

It does not matter whether the Photon is plugged into the breadboard or not. However, keeping your Photon on the bread-board will protect its pins against getting bent or accidentally shorting together if they touch something metal.

Error Messages

If you mistyped something, you will see an error message. A common mistake is a missing semicolon on the end of a line. If this is your error, then if you look carefully at the error message, you will see a lot of incomprehensible stuff before you get to this line:

```
the_user_app.cpp:3:1: error: expected ';' before '}' token
```

This is just saying that there is a semicolon missing somewhere. So go back and check over your code carefully.

Once the Photon/Core has finished flashing purple and has restarted, you should see the RGB LED cycle through red, green, and blue.

Summary

Congratulations, you have made your Photon or Core do something! In the next chapter, you will revisit both the apps that you have run on your Core to understand a bit more about how they work and how to go about writing your own apps.

3/Programming the Photon

Chapter 2 gave you your first taste of programming the Photon/ Core. This chapter will take you deeper into the world of programming and, if you are new to coding, teach you some of the basics of the C language that these devices use.

If you are a seasoned programmer, you can skip big chunks of this chapter.

The Web IDE

In Chapter 2, you touched on using the Web IDE.

Apart from typing in the code and clicking a couple of the buttons, there are large uncharted areas of the Web IDE to explore. Figure 3-1 shows the Web IDE with labels for some of its buttons.

You have already used the Flash button at the top of the button panel. The Verify button will check a program and make sure the code is legal without actually sending it to the Photon. Since the Flash button also verifies the program and will not send it unless it's legal, you may find the Verify button to be of limited use.

The Save button will save whatever app is currently showing in the editor window to the right. The Code button toggles the side panel that contains your list of apps and the buttons that allow you to remove an app or create a new app, as you did in Chapter 2.

Flash

Verify

Save

Code

Libraries

Docs

Devices

Settings

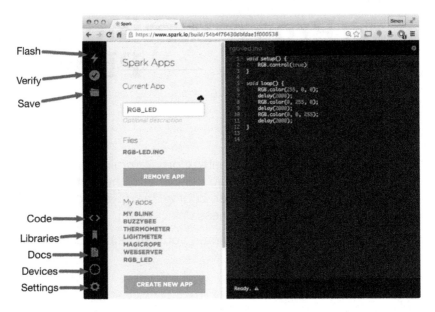

Figure 3-1. *The Web IDE*

Underneath the Code button, the Libraries button allows you access to a huge body of community-contributed libraries of code that you can make use of in your apps. Many of these libraries are concerned with interfacing specific types of hardware such as sensors and displays.

The Docs button opens up the web page for the Photon's documentation.

The Devices button allows you to manage your Photons and Cores. When you click it, a side panel slides out, listing all the devices associated with your account so that you can manage them in various ways (Figure 3-2).

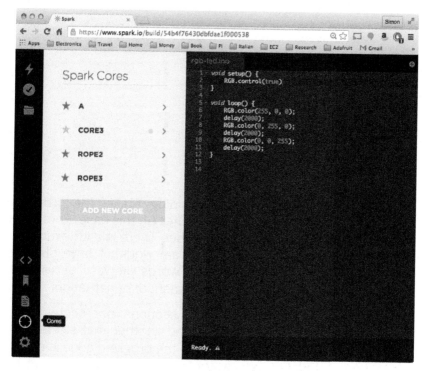

Figure 3-2. *Managing your Photons*

The list will show all the Photons and Cores you have that are associated with your account. The gold star next to the device indicates that this is the currently selected device, so if you clicked the Flash button, this is the device that would be programmed.

The blue dot to the right of the device name indicates that the device is active and connected to the cloud service.

Coding an App

Let's start by dissecting the app that you wrote in Project 2, which changed the color displayed on the Photon's built-in LED to red, green, or blue at two-second intervals. As a reminder, the code for this is listed here:

```
void setup() {
    RGB.control(true);
}

void loop() {
    RGB.color(255, 0, 0);
    delay(2000);
    RGB.color(0, 255, 0);
    delay(2000);
    RGB.color(0, 0, 255);
    delay(2000);
}
```

A program (or *app* in Particle terminology) takes the form of a series of instructions that the computer (in this case, Photon or Core) must execute one after the other. There is a little more to it than this, or the preceding program wouldn't have strange { and } symbols as well as mystical words like **void**, **setup**, and **loop** and a seemingly random assortment of punctuation.

As you can see, the code is in two sections: one section that starts with the words **void setup** and another that starts with **void loop**. Each of these sections of the program code is called a *function*. A function is a way of grouping together a set of instructions and giving them a name.

Looking at the first function, we see it begins with the word **void**. This might seem an unusual word to indicate the start of a function. In case you are interested, this is because every function has to specify a *type* (more on types later). Both the functions in this program have no type, so we have to indicate this fact using the word **void**. One thing that you will notice right away about programming languages is that they are really fussy. Computer scientists love things to be orderly and precise; hence we need to write **void** before our **setup** and **loop** functions.

After the word **void** comes the function name **setup** followed by (). Later on, you will see functions that have something inside the () called *parameters*. Again, the precise nature of the C language dictates that even if this function happens not to have any parameters, you must still include the parentheses.

Every app must include the functions **setup** and **loop**. That is why when you click the CREATE NEW APP button, the skeleton code for these two functions will appear as follows, ready for you to add in your own instructions:

```
void setup() {
}

void loop() {
}
```

The program code that you add to a function needs to go between the { and }, with a ; after each instruction line. Over the years, programmers have developed conventions on how you should write your code to make it as easy as possible for another programmer to understand, or for you to understand it when you return to it after not looking at it for a while.

These conventions include indenting each of the lines inside the function, so that it is easy to see that they belong to it. You will hear such sections of indented code being referred to as a *block* of code. Unlike the Python language, the C language used by the Photon and Core ignores all indentation. If you leave out the indentation, the program will still work just fine; it's just a matter of convention to keep things neat.

The use of the two functions **setup** and **loop** is borrowed from Arduino and works in just the same way. When your device is first powered up, or after you press the Reset button, the lines of code inside the **setup** function will be *run* (executed) just once. They will not be run again until the Photon or Core is reset.

In the case of our example program, the **setup** function has just a single line of code in it:

```
RGB.control(true);
```

This command tells the Photon/Core that we want to control the color of its RGB LED rather than let it do all its fancy blinking to tell us its status. Don't worry, the device will revert to blinking its status colors when we upload a different app that does not take control of the RGB LED. We will still be able to send new apps to the the Photon/Core; we just won't know whether it is connected properly to the Internet with the app running.

The **loop** function is different from the **setup** function because it will run over and over again. As soon as it has finished running the last of the commands within its curly braces, it will start running the first one again.

In this case, there are four lines of code that will be run each time around the loop. The first line is as follows:

```
RGB.color(255, 0, 0);
```

Our setup function code gave us control of the RGB LED. Now we need to instruct it to change color. The first line of the loop code sets the color to red. It does this because RGB.color is a function like setup and loop, but in this case, a function that is built into the Photon's software that we can make use of. Whereas the setup and loop functions had empty () after them, you can see that RGB.color has three parameters, separated by commas. These three parameters represent the brightness of the red, green, and blue channels of the LED, respectively. The number supplied for each parameter is between 0 and 255, where 0 means no light, and 255 means full brightness. So, the values of 255, 0, 0 mean full brightness red, no green, and no blue. If we were to use 255, 255, 255, the LED would shine all three colors at full brightness and appear white.

Once the color is set to red, we want the app to do nothing for a couple of seconds before we change the color to green. There is another built-in function that we can use for that called delay. So we use this in the second instruction of the loop function:

```
delay(2000);
```

This causes the Photon/Core to twiddle its virtual fingers for 2000 milliseconds. A millisecond is 1/1000 of a second, so 2000 of them is the same as 2 seconds.

The first two lines of the loop function are repeated two more times, to change the LED color to green and then to blue:

```
void loop() {
    RGB.color(255, 0, 0);
    delay(2000);
    RGB.color(0, 255, 0);
    delay(2000);
    RGB.color(0, 0, 255);
    delay(2000);
}
```

You might be wondering if you need the final **delay** command inside **loop**. Try deleting it and then flashing the code to your Photon/Core again. The updated program will not turn the LED blue at all. In fact, what is happening here is that the LED is being set to blue, but then faster than the human eye, the app is back around to the first line of **loop** again, and the LED will be set to red.

Each time around the loop, the Photon/Core will also do a quick check to see if the cloud service is trying to push it a new app. If there is a new app, then your app will be interrupted, and the Photon/Core will take back control of the LED and start flashing purple to indicate that the new app is being "flashed" onto the Photon/Core.

Comments

Project 1 was simply to blink the blue LED connected to D7 by using the example program provided by Particle. It's actually a more complex program than that of Project 2, so let's build on what you have seen in Project 2 and now work through Project 1. Along the way, you will encounter various programming ideas.

It will help to have the code for the example app "Blink an LED" open in the code editor of the Web IDE as the code is examined. Here's the code. It gets pretty wide, so the righthand side of the text has been truncated in the following listing:

```
// Define the pins we're going to call pinMode on
int led = D0;  // You'll need to wire an LED to this one to
               // see it blink.
int led2 = D7; // This one is the built-in tiny one to the
               // right of the USB jack

// This routine runs only once upon reset
void setup() {
  // Initialize D0 + D7 pin as output
  // It's important you do this here, inside the setup()
  // function rather than outside it or in the loop()
  // function.
  pinMode(led, OUTPUT);
  pinMode(led2, OUTPUT);
}
```

```
// This routine gets called repeatedly, once every 5-15
// milliseconds.
// Spark firmware interleaves background CPU activity
// associated with WiFi + Cloud activity with your code.
// Make sure none of your code delays or blocks for too long
// (more than 5 seconds), or weird things can happen.
void loop() {
  digitalWrite(led, HIGH);    // Turn ON the LED pins
  digitalWrite(led2, HIGH);
  delay(1000);                // Wait for 1000mS = 1 second
  digitalWrite(led, LOW);     // Turn OFF the LED pins
  digitalWrite(led2, LOW);
  delay(1000);                // Wait for 1 second in off mode
}
```

The first thing to notice is that the program has a whole load of lines that start with //. Starting a line with // indicates that the line is not program code at all, but a comment that tells the reader something *about* the code. The following line tells us pretty much all we need to know about what the next couple of lines are for:

```
// Define the pins we're going to call pinMode on
```

The two lines that follow are a mixture of real code and comment. The // is used again after the semicolon to indicate that the rest of the line is also a comment:

```
int led = D0;   // You'll need to wire an LED to this one to
                //see it blink.
int led2 = D7;  // This one is the built-in tiny one to the
                // right of the USB jack
```

The commenting in this particular app is really quite excessive, and it is deliberately very detailed to help the novice try to work out what on earth is going on with the program.

Variables

The two preceding lines define something called *variables*. Variables are an incredibly useful programming concept. They allow you to give something a meaningful name.

In the case of the first of the lines, there is a pin named D0. That's fine. We know that this refers to one of the physical pins on the Photon/Core that can be used as an input or output, but it says absolutely nothing about what that pin is to be used for in the rest of the app. Actually, as the comment says, an LED is going to be attached to D0 (you will do this in Chapter 4). Later in the code, when we want to turn the pin on and off, it's nicer to refer to that pin as the LED pin rather than the actual pin number. The following command is said to assign the value D0 to the variable named `led`:

```
int led = D0;
```

Once the variable has been assigned the value D0, from then on we can refer to pin D0 as `led` rather than D0. The word `int` at the start specifies the *type* of the variable. The type `int` means a number, and while you might argue that D0 does not look like a number, actually, behind the scenes, D0 is a number used internally by the Photon/Core.

In a later section, we will explore the concept of variable types.

Now that a variable called `led` has been defined to refer to D0, you can use that variable both in the `setup` function, where we set the `pinMode` of the LED pin, and in the `loop` function, where it is turned on and off to make it blink.

Using a variable like this has the great advantage that, if you decided to use pin D3 instead of D0, then you would have to change only the single line

```
int led = D0;
```

to

```
int led = D3;
```

Morse Code

Now, you might wonder why a book about the latest IoT hardware like the Photon would be talking about a 19th century invention called Morse code. The answer lies in the fact that there are only so many things you can do with an LED, and flashing out Morse code signals is one of the more interesting ones.

Morse code represents letters and numbers as short (dot) or long (dash) beeps or flashes of light. For example, the letter *A* is represented as .- (dot dash), one short pulse followed by one long pulse. Although there are a few variants of Morse code, we will use International Morse Code here.

If you were to convert the word PHOTON to Morse code, the resulting message would be .--. .... --- - --- -. Morse code does not distinguish between uppercase and lowercase letters.

International Morse Code

The following table lists the International Morse code letters and digits:

A	.-	N	-.	0	-----
B	-...	O	---	1	.----
C	-.-.	P	.--.	2	..---
D	-..	Q	--.-	3	...--
E	.	R	.-.	4	....-
F	..-.	S	...	5	.....
G	--.	T	-	6	-....
H	....	U	..-	7	--...
I	..	V	...-	8	---..
J	.---	W	.--	9	----.
K	-.-	X	-..-		
L	.-..	Y	-.--		
M	--	Z	--..		

Flashing SOS

SOS is ... --- ... in Morse code and is used by someone in distress, usually on the high seas and usually a hundred years ago.

We have already made an LED blink, so let's look at how we can apply what we have learned to flashing a distress signal (albeit, on the small RGB LED).

As a reminder, here is the code from Project 2:

```
void setup() {
    RGB.control(true);
}

void loop() {
    RGB.color(255, 0, 0);
    delay(2000);
    RGB.color(0, 255, 0);
    delay(2000);
    RGB.color(0, 0, 255);
    delay(2000);
}
```

 Getting the Code

If you typed this code in and got it working in Chapter 2, all is well and good. If not, the easiest way to get this into the IDE and then onto your Photon/Core is to click the Libraries button in the Web IDE and find the library called *PHOTON_BOOK*. You will find all the project programs used in this book as examples.

Select the file *p_02_RGB_LED* and then click the button USE THIS EXAMPLE. A copy of the program will open up in the editor, and you can then flash it to your Photon/Core. Note that this will be your own copy of the original program, so you can modify it if you wish.

Flash the program onto your Photon/Core, just to remind yourself of the process and see what it does.

While multicolor Morse code could be useful, we should just start by doing some blinking in white light. Change the code of Project 2 to simply blink white (255, 255, 255), as shown here:

```
void setup() {
    RGB.control(true);
}

void loop() {
    RGB.color(255, 255, 255);
    delay(500);
    RGB.color(0, 0, 0);
```

```
    delay(500);
  }
```

You can also find this code in the examples library as *ch_03_blink_white*.

The sequence of instructions in **loop** is now as follows:

1. Set the RGB LED to white (red, green, and blue, all maximum brightness).
2. Delay half a second.
3. Turn off the RGB LED.
4. Delay half a second.
5. Start again.

Flash the app onto your Photon/Core and you should see the LED blink white once per second.

To flash SOS (... --- ...), we can just add some more lines to **loop** and vary the delays. A dash is three times as long as a dot, and the gap between one dot or dash and the next should be the same as a dot. The gap between one letter and the next should be the same as a dash, so change the program so that it appears as follows. Note that using copy and paste will save you some time.

```
void setup() {
    RGB.control(true);
}

void loop() {
    // flash S
    // dot
    RGB.color(255, 255, 255);
    delay(200);
    RGB.color(0, 0, 0);
    delay(200);
    // dot
    RGB.color(255, 255, 255);
    delay(200);
    RGB.color(0, 0, 0);
    delay(200);
    // dot
    RGB.color(255, 255, 255);
    delay(200);
    RGB.color(0, 0, 0);
```

```
delay(600);
// end of S
// flash O
// dash
RGB.color(255, 255, 255);
delay(600);
RGB.color(0, 0, 0);
delay(200);
// dash
RGB.color(255, 255, 255);
delay(600);
RGB.color(0, 0, 0);
delay(200);
// dash
RGB.color(255, 255, 255);
delay(600);
RGB.color(0, 0, 0);
delay(600);
// end of O
// flash S
// dot
RGB.color(255, 255, 255);
delay(200);
RGB.color(0, 0, 0);
delay(200);
// dot
RGB.color(255, 255, 255);
delay(200);
RGB.color(0, 0, 0);
delay(200);
// dot
RGB.color(255, 255, 255);
delay(200);
RGB.color(0, 0, 0);
delay(600);
// end of S

delay(2000); // delay before repeat
}
```

While this code works just fine, there is much more of it than there needs to be, as a lot of the code is repeated. In the next section, you will use functions to shorten the code considerably.

Functions

We have been happily using some of the functions such as delay and RGB.color that are built in to the Spark.io system. You can also write your own functions to group lines of code that you might want to use over and over again in your program.

For example, wouldn't it be great if there was a function called flash, by which we could just pass as parameters the duration of the flash and how long to pause before any further flashing? Well, you can create such a function and then modify the SOS program to use it. The function looks like this:

```
void flash(int duration, int gap) {
    RGB.color(255, 255, 255);
    delay(duration);
    RGB.color(0, 0, 0);
    delay(gap);
}
```

The function starts with the word **void**, because this function does not return any value. Next we have the name of the function, and I have decided to call it **flash**. Inside the parentheses two parameters are defined, **duration** and **gap**. Both need to have the type specified for them. Since they are a whole number of milliseconds, the type **int** should be used.

The body of the function is just like a section of the long-winded code we had before, but now, instead of having fixed numbers like 200 or 600 inside the **delay** functions, we have what look like variable names of **duration** and **gap**. These values will be passed into the function when we use it in **loop**.

To see how this works, let's now look at the whole program, which you can find in *ch_03_SOS_function*:

```
void setup() {
    RGB.control(true);
}

void loop() {
    // flash S
    flash(200, 200);
    flash(200, 200);
    flash(200, 600);
```

```
        // flash O
        flash(600, 200);
        flash(600, 200);
        flash(600, 600);
        // flash S
        flash(200, 200);
        flash(200, 200);
        flash(200, 600);

        delay(2000); // delay before repeat
    }

    void flash(int duration, int gap) {
        RGB.color(255, 255, 255);
        delay(duration);
        RGB.color(0, 0, 0);
        delay(gap);
    }
```

This is great because the program is just 27 lines long rather than the 59 lines of the previous version. Smaller programs are better, being easy to maintain. Programmers have an acronym that they use—DRY, for Don't Repeat Yourself—and will talk of good code being DRY.

Now, to flash the letter S, you can just call flash three times in a row. On the last occasion, a longer delay is used to allow a gap between letters.

We could take this program a step further and create two new functions called flashS and flashO that call flash. This would simplify the loop code even more. The result of this can be found in the file ch_03_SOS_function2.

```
    void setup() {
        RGB.control(true);
    }

    void loop() {
        flashS();
        flashO();
        flashS();

        delay(2000); // delay before repeat
    }
```

```
void flash(int duration, int gap) {
    RGB.color(255, 255, 255);
    delay(duration);
    RGB.color(0, 0, 0);
    delay(gap);
}

void flashS() {
    flash(200, 200);
    flash(200, 200);
    flash(200, 600);
}

void flashO() {
    flash(600, 200);
    flash(600, 200);
    flash(600, 600);
}
```

This hasn't made the program any shorter, but an interesting thing has happened. Some of the comments in loop have become so obvious that they can now be removed. For example, there is little point in preceding the line

```
flashS();
```

with a comment like

```
// flash S
```

Generally, the fewer comments that are needed to "explain" the code, the better. The code has become less obscure and is starting to be self-explanatory.

Incidentally, it does not matter where in the program you write the code for a function. However, the convention is to have the setup and loop functions at the top of the file, as these are the root of any later calls to other functions. In other words, setup and loop will always be the starting point for anyone reading through the program to work out what it does.

Types

The SOS program is now looking pretty neat. You can start at loop, and understand that loop is going to call flashS and flash0. You can then look at flashS and see that flashS calls flash. It's all quite easy to "read." The program does, however, have one weakness: should you want to make it flash SOS faster or slower, you would have to go through and change all occurrences of 200 and 600 to something else. It's easy to fix this so that you have to change only one variable.

The modified version can be found in the file *ch_03_sos_vars*.

```
int dot = 200;
int dash = dot * 3;

void setup() {
    RGB.control(true);
}

void loop() {
    flashS();
    flash0();
    flashS();

    delay(2000); // delay before repeat
}

void flash(int duration, int gap) {
    RGB.color(255, 255, 255);
    delay(duration);
    RGB.color(0, 0, 0);
    delay(gap);
}

void flashS() {
    flash(dot, dot);
    flash(dot, dot);
    flash(dot, dash);
}

void flash0() {
    flash(dash, dot);
    flash(dash, dot);
    flash(dash, dash);
}
```

At the top of the file, two new variables, **dot** and **dash,** have been added. The variable **dot** is set to 100. This will be the duration of the delay for a dot flash. The variable **dash** is the duration of a dash, and we could just set this directly to 600, but then if we wanted to make the SOS blink faster or slower, we would have to change both variables. Since a dash is always three times as long as a dot, we can express that in our code like this:

```
int dash = dot * 3;
```

Here, you are defining a new variable called **dash** (of type **int**) and then assigning it a value of **dot** * 3. In the C language, * means *multiply*.

As well as * for multiply, you can use + for add, - for subtract, and / for divide.

Change **dot** to **100** and then flash it onto your Photon/Core. Notice how the blinking is now faster.

The int Type

We have established that the type **int** is used for whole numbers. These numbers can be positive or negative or 0 but must be in the range −2,147,483,648 to 2,147,483,647.

Note that if you have come from the world of Arduino, where the range of numbers in an **int** is much smaller, this will be a pleasant surprise for you. Although C has other integer types like **long** (same range as **int**) as well as signed and unsigned versions of the types, there is little point in ever using anything other than **int** to represent whole numbers.

The float Type

Variables of type **int** are great for whole numbers, like pin numbers or counting, but, occasionally, you might want to use numbers with a decimal place in them. For example, if you are reading temperatures from a sensor, the nearest degree might not be precise enough, and you might want to represent a value of, say, 68.3 degrees.

For these kinds of numbers, you cannot use **int**. You use a type called *float*. They are called *floats*, because they are *floating-point* numbers. That is, the position of the decimal place could be anywhere in the number; they are not, say, fixed to two decimal places.

Here is an example of a float:

```
float temperature = 68.3;
```

Floats have a vast range, from −3.4028235E+38 to 3.4028235E+38. The notation E+38 means with 38 zeros on the end of the number. These numbers are as big as you could ever need. For the mathematically minded, floats here are 32 bits.

So, if floats have a much wider range, why bother with restricted int values? Surely we could just use floats, and if they need to be whole numbers, just have .0 on the end.

The reason this is not done is that floats are deceptive. Although they have a wide range of possible values, they do this by using an internal representation that effectively approximates the number. Although they represent numbers that can be 39 digits long, if you performed some arithmetic on two numbers that were that long, the result would not be exact and under some circumstances, you can get results equivalent to 2 + 2 = 3.9999999999999999999.

The likelihood of such errors can be reduced by using the type double, which uses a 64-bit representation and is, therefore, a lot more precise.

Arithmetic is slower when using float than when using int, and, if you use double, it takes up twice as much memory. So, avoid using float or double unless you have a good reason for it, such as the temperature example.

Other Types

As this book progresses, you will meet other data types such as boolean, which represents the values true and false, as well as more complex types, like string, which is used to represent text.

Arrays

The variables used so far have all been just single values. Sometimes you need to use a data structure that represents a list of values—for example, a list of flash delays.

Open the app *ch_03_SOS_function* in the Web IDE. This is taking a step backward, but this version is quite suitable to modify to use an array of delay values.

The `loop` function of this program looks like this:

```
void loop() {
    // flash S
    flash(200, 200);
    flash(200, 200);
    flash(200, 600);
    // flash O
    flash(600, 200);
    flash(600, 200);
    flash(600, 600);
    // flash S
    flash(200, 200);
    flash(200, 200);
    flash(200, 600);

    delay(2000); // delay before repeat
}
```

If you removed the comments from this code, you would have nine flash commands one after the other, each with a series of values both for duration and gap. We found one way to simplify this code using functions called **flashS** and **flashO**, but another way to simplify the code would be to use two arrays: one for the duration and one for the gap. We could then go through each position of these arrays in turn, calling **flash** with the values. That way, we could change the Morse code message to anything we liked, just by changing the contents of the arrays.

This is how you define an array of *ints*:

```
int durations[] = {200, 200, 200, 600, 600, 600, 200, 200,
200};
```

Notice that the variable name now has [ ] after it to indicate that it is an array rather than a single value. The values separated by commas inside { and } are the values to be put into the array. In this case, that is each of the durations used in `loop`.

We could do the same for the gap parameters of the calls to `flash`, like this:

```
int gaps[] = {200, 200, 600, 200, 200, 600, 200, 200, 600};
```

It's almost like there are two tapes with the same numbers on them that are then going to be fed through a machine that will use one number for the flash duration and the other for the gap before moving to the next position on the tape.

Loops

Now that we have the two arrays, we need a way to step over each of those values in turn and call **flash** using them.

This type of thing is called *looping* or *iteration*, and there is a useful language command in C for doing this called **for**.

Here is the modified program that uses **for** to loop over each element of the arrays and flash them. If you want to try out the completed program, it's in *ch_03_SOS_Array*.

```
int durations[] = {200, 200, 200, 600, 600, 600, 200, 200,
200};
int gaps[] =      {200, 200, 600, 200, 200, 600, 200, 200,
600};

void setup() {
    RGB.control(true);
}

void loop() {
    for (int i = 0; i < 9; i++) {
        flash(durations[i], gaps[i]);
    }

    delay(2000); // delay before repeat
}

void flash(int duration, int gap) {
    RGB.color(255, 255, 255);
    delay(duration);
    RGB.color(0, 0, 0);
    delay(gap);
}
```

The syntax of this **for** command is confusing. It looks a bit like a function call, but it isn't, because it has parts separated by semicolons. Surely, these should be commas?

If you are new to programming, it's probably best for you to just use the **for** loop for counting, copy the line starting with **for**, and change 9 to whatever number you want i to count up to. It's not you; the syntax really is a bit screwy.

What is going on is that the first thing inside the parentheses of the **for** command defines a variable. In this case, the variable is called i (for *index*), and this variable is initialized to 0 and will be used to keep track of our position in the arrays.

Then there is a semicolon and the expression $i < 9$. This is the condition for staying in the loop. In other words, the program will not escape from **for** and go on to do the last **delay** command while i is less than 9. This means that our program will hang unless we do something to change the value of i. That is where the last part of the stuff inside the () of the **for** command comes into effect. The expression i++ means add 1 to the value of i.

So, each time around the **for** loop, **flash** will be called with the *ith* value of durations (that's what [i] means) and the *ith* value of gaps. The variable i will then have 1 added to it; then **for** will call **flash** again, but this time i will have changed from 0 to 1. This will continue until i gets to 9, and then because i is now greater than 8, the **for** loop is finished and the program will move on to the last line of the **loop** function and delay for two seconds.

As we mentioned earlier, if we change what is in the arrays, we can change the message completely. To try this, load the example *ch_03_flash_photon*. Note that the maximum value of i in the **for** loop has been changed from 9 to 16 because the arrays have gotten bigger.

You may have noticed that i starts at 0 rather than 1. In C, the first element of an array is the *zeroth* element rather than the first element.

Strings

The Morse code example is gradually evolving away from something that very specifically can flash only SOS into something that will eventually be able to flash any message that we tell it to.

To do this, we need a way for the program to represent text. For this, we will introduce a new type called `String`. Here is how you define a `String` variable:

```
String message = "My Photon speaks Morse";
```

The value given to the string is enclosed in double quotes. If you are used to languages like Python that will allow you to use double or single quotes, then beware, because only double quotes will do for C. C does use single quotes, but these are reserved for single characters of type `char`. You will meet the `char` type a little later on.

There are various things that you can do with strings. For a start, you can find out how many characters they contain, like this:

```
String message = "My Photon speaks Morse";
int len = message.length();
```

The function `length` is a special kind of function called a *method*. The difference between a method and a function is that a method is owned by a type, so here strings have a method called `length` that you can call using the dot notation after the string variable name. In this case, the `length` method returns a value (the number of characters in the string) that can then be assigned to an `int` variable.

As we mentioned earlier, C has a type specifically for single characters. Since a string is made up of a sequence of characters, it's a bit like an array. You can access any of the string's individual characters by using the method `charAt()`. The following example will assign the value *P* to the variable `letter`:

```
String message = "My Photon speaks Morse";
char letter = message.charAt(3);
```

Just like arrays, the index positions start at 0 rather than 1.

There are lots of other things that you can do with strings, and we will encounter these as we finish the Morse Code Flasher project.

Ifs

The C if command is one that you will find in pretty much any program of any size. It allows the program to make decisions as to what to do based on some condition. It works much the same way as the English word *if* in a sentence. So, the English statement "if the character is a space, delay for seven dots' worth of time" (the gap between words in Morse code is seven times the length of a dot) would be written in code as follows:

```
if (letter == ' ') {
    delay(dot * 7);
}
```

Notice how, when comparing a letter to a space character, the two equals signs are used. This is to distinguish it from the single equals sign used to give a variable a value. A common source of programs not doing what was expected is that = was used instead of ==.

You can use == to test if two things are equal, or != to test that they are not equal. You can also use < (less than), <= (less than or equal to), > (greater than), or >= (greater than or equal to).

It is also possible to combine more than one condition by using && (and) and || (or). For example, the block of code inside this if statement will be run only if the value of letter is between *A* and *Z*. You can use these comparisons on ints, floats, and chars, but not strings.

```
if (letter >= 'a' && letter <= 'z') {
    // it's a lowercase letter do something!
}
```

The else command works with if. The usage is the same as in English. For example, in English you could write, "If the temperature is less than 70, turn on the heating, or else turn it off."

In C, this might look like this:

```
if (temperature < 70.0) {
    heatingOn();
```

```
else {
    heatingOff();
}
```

I have assumed that two C functions called `heatingOn` and `hea`
`tingOff` exist to control the heating.

Project 3. Morse Code Flasher

In this project, you will put together everything you have learned
about programming the Photon/Core. This project will flash any
plain text held in a string as Morse code.

In Project 4, you will use this same code with an external LED
and in Project 8 flash Morse code messages using a web interface.

Software

You can find the code for this project in the example library as
p_03_Morse_Flasher. I suggest you open this file up in the Web
IDE while you are looking at the code.

The first line of code specifies the message that is to be flashed:

```
String message = "My Photon speaks Morse";
```

You can change this message to anything you like.

Following this are the variables **dot** and **dash** that set the timings
for the flashes. Then we have an array of strings that defines all
the letters in Morse code:

```
int dot = 200;
int dash = dot * 3;

String letters[] = {
  ".-",  "-...",  "-.-.",  "-..",  ".",  "..-.",  "--.",  // A-G
  "....", "..",  ".---",  "-.-",  ".-..",  "--",  "-.",  // H-N
  "---",  ".--.",  "--.-",  ".-.",  "...",  "-",  "..-",  // O-U
  "...-", ".--",  "-..-",  "-.--",  "--.."  // V-Z
};
```

Each of the letters *A* to *Z* is represented by one element of the
array. The first element of the array is the string ".-" for *A*, the
second "-..." for *B*, etc. This array will be used to look up the

series of dots and dashes that must be flashed in order to flash a particular letter as Morse code.

The `setup` function takes control of the RGB LED, and the `loop` function now contains the function call to `flashMessage`.

```
void setup() {
    RGB.control(true);
}

void loop() {
    flashMessage(message);
    delay(5000); // delay before repeat
}
```

This type of programming is called *programming by intention*. You know what you intend to do in the `loop` function: you want to flash the message and then delay for a while before flashing the whole message again. You can worry about how the `flashMes sage` function will actually do its job later.

In fact, implementing `flashMessage` is the next step. Logically, to be able to flash a whole message you need to flash each letter of that message in turn:

```
void flashMessage(String message) {
    for (int i = 0; i < message.length(); i++) {
        char letter = message.charAt(i);
        flashLetter(letter);
    }
}
```

So, all `flashMessage` needs to do is to step over each letter in turn and call a function `flashLetter` on that letter. Once again, we are deferring the actual LED flashing work to another function that we have not written yet, called `flashLetter`. See how nicely the problem is breaking down for us?

```
void flashLetter(char letter)
{
    if (letter >= 'a' && letter <= 'z') {
        flashDotsAndDashes(letters[letter - 'a']);
    }
    else if (letter >= 'A' && letter <= 'Z') {
        flashDotsAndDashes(letters[letter - 'A']);
    }
    else if (letter == ' ') {
```

```
        delay(dot * 7); // gap between words
    }
}
```

The `flashLetter` function is a little more complicated. First we need to use `if` commands to decide if the letter is an uppercase letter, a lowercase letter, or a space character. If it's a lowercase letter, we find the string containing the dots and dashes for that letter, using the `letters` array and supplying the character code minus the code for *a* to get the right sequence of dots and dashes from the `letters` array.

If the letter is uppercase, you need to subtract the character code for *A* rather than *a*.

Finally, if the letter is the space character, that indicates that it's the end of a word and the Morse code standard is to leave a gap of seven dots' worth of time between each word.

Yet again, we are deferring the job of making the flashes to a function called `flashDotAndDashes`:

```
void flashDotsAndDashes(String dotsAndDashes) {
    for (int i = 0; i < dotsAndDashes.length(); i++) {
        char dotOrDash = dotsAndDashes.charAt(i);
        if (dotOrDash == '.') {
            flash(dot);
        }
        else {
            flash(dash);
        }
        delay(dot); // gap between dots and dashes of a letter
    }
    delay(dash - dot); // gap between letters of a word
}
```

The parameter passed to `flashDotAndDashes` is a string. This string will be made up of the dots and dashes for a particular letter (for example, -... for *B*). So to flash out those dots and dashes, you need to step over each character in that string, flashing the dot or dash.

After you have flashed each dot or dash, you need one dot's worth of delay before you start on the next dot or dash. When you have completed the whole sequence, you need another delay of duration dash. However, you will already have had a delay of one dot after the last dot or dash was flashed, so the delay is for the duration of a dash minus a dot.

And still we are not actually controlling the LED. This happens in the `flash` function, which just takes the duration of the flash as a parameter. It is this function that turns the LED on and off:

```
void flash(int duration) {
    RGB.color(255, 255, 255);
    delay(duration);
    RGB.color(0, 0, 0);
}
```

That is all there is to this project. Try changing the message and also the speed of the Morse code by altering the value of the **dot** variable.

Summary

This chapter has only really touched on programming the Photon/Core, providing enough to get you started. As the book progresses, you will learn more about programming the Photon.

In the next chapter, you will learn how to connect hardware to the Photon/Core by using a solderless breadboard.

4/Breadboard

In this chapter, we will deal with attaching some simple external electronics to the Photon or Core, including LEDs and switches. To connect electronic components like LEDs to the Photon/Core, you will use something called a *solderless breadboard*.

The solderless breadboard (or just *breadboard*) was invented as a tool for electronics engineers to prototype their designs before committing the projects to a more permanent soldered form. They do, however, make an excellent tool for you to experiment with electronics and make your own projects without the need for any soldering.

Figure 4-1 shows a breadboard with a Spark Core, an LED, and a resistor attached to it. This could also be a Photon.

When you buy a Photon or Core, one of the options available to you is for it to include a breadboard and header pins. If you did not select this option but rather bought a bare Photon, don't worry; you will find sources for the header pins and breadboard that you will need in Appendix A (it also includes sources of components for the projects that follow).

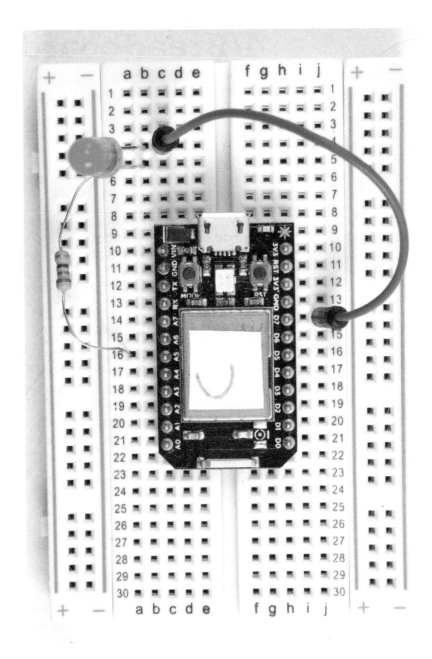

Figure 4-1. *A Core on a breadboard*

How a Breadboard Works

Behind the holes in the front plastic face of a breadboard, there are metal clips designed to grip wires and component leads. These clips are arranged into rows of five, in two banks of 30 rows. At least, they are in the fairly standard half-sized breadboard that is included as an option when you buy a Photon.

Down each side of the breadboard are pairs of long clips that are marked with red and blue lines on the front of the breadboard. These are often used when connecting power to various parts of your breadboard design.

Figure 4-2 shows a breadboard that has been taken apart and one of the clips removed, so that you can see how it works behind the plastic.

Figure 4-2. *A breadboard disassembled*

Attaching an LED

LEDs emit light when a current passes through them. Unfortunately, LEDs are kind of greedy for current and have so little restraint that, when fed with a source of voltage like an output pin on a Photon/Core, they will draw as much current as they can. This can lead to their overheating, shortening their life, or even damaging the Photon.

To prevent such problems, LEDs are used with the electronic equivalent of a gastric band that restricts the flow of current into the LED. The component that restricts the flow of current in this way is called a *resistor*.

Figure 4-3 shows a schematic diagram for using a resistor with an LED.

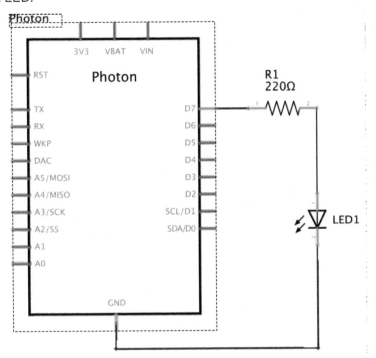

Figure 4-3. *Limiting the current to an LED with a resistor*

The two pins of the Photon that are used are GND and D7. GND, or ground, is the base voltage against which any other voltages on the Photon are measured. For example the 3V3 pin is 3.3V above (more positive than) GND. Current will flow only from a higher voltage to a lower one, so if the digital output D7 is at 0V (called LOW), no current will flow, but if it is at 3.3V (HIGH), then the current can flow out of D7 through the resistor (zigzag line), through the LED, and down to GND. You may find that it helps to think of current flowing like water, from a higher elevation down to sea level.

Digital Outputs

The pins on your Photon/Core labelled D0 to D7, as well as the pins A0 to A5, can all be used as digital outputs. That is, you can write some commands in your program that will turn them on or off. More accurately, the pins are set to either 3.3V or 0V (GND).

Because the pins can be used as either inputs or outputs, the first thing you need to do in your **setup** function, for all the pins you are using, is to specify whether that pin is to be used as an input or an output. In "Digital Inputs" on page 68, you will find out how to specify a pin as being an input. To specify that a pin is going to be used as an output, include the following command in your **setup** function

```
pinMode(pin, OUTPUT);
```

When it comes to setting the pin HIGH or LOW, you use the command **digitalWrite**. So, assuming that the variable **pin** is set to one of the pins (say, D7), then to set the pin HIGH, you would use

```
digitalWrite(pin, HIGH);
```

and to set it low, you would use

```
digitalWrite(pin, LOW);
```

In Project 4, you will use these commands to make an external LED blink Morse code.

Project 4. Morse Flasher (External LED)

Now that you know how to use an LED with a Photon or Core, you can change the Morse code flasher of Project 3 to use an external LED rather than the built-in RGB LED.

Parts

To build this project, you need the parts listed in Table 4-1 in addition to your Photon or Core.

Table 4-1. *Project 4 parts bin*

Part	Description	Appendix A code
LED1	Red, or for that matter, any other color LED	C2
R1	220Ω resistor	C1
	Male-to-male jumper wire	H4
	Half-sized breadboard	H5

Hardware

We can convert the schematic of Figure 4-3 into a breadboard layout, as shown in Figure 4-4.

Start by pushing one end of the resistor (it does not matter which) into the same row as D7 on the Photon/Core and then put the other leg of the resistor into row 4 of the breadboard.

The LED will have one leg longer than the other. The longer leg is the positive leg, and this is the leg that should go to the same row as the resistor.

Finally, connect a jumper wire from the row of the breadboard on the lefthand side, which is connected to one of the Photon/Core GND pins, to the top row that the LED is connected to.

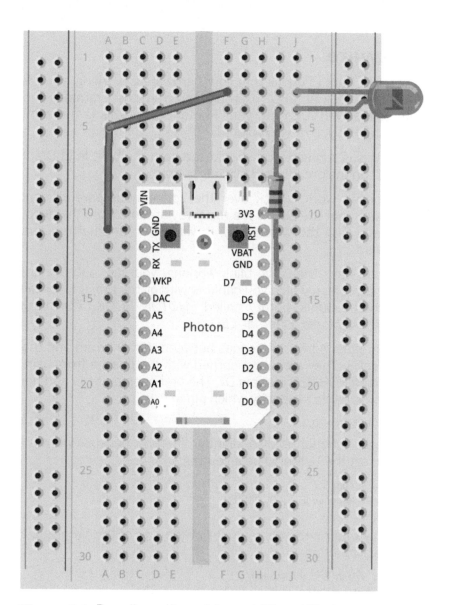

Figure 4-4. *Breadboard layout for an LED and Photon*

Software

If you just want to check that your hardware is working, you can flash the "Blink an LED" example app from the Examples section of the Web IDE onto your Photon/Core. This is the same program that you ran in Project 1.

When this is running, you should see both the built-in LED and the LED on the breadboard blinking in time.

If it doesn't work, check over the connections carefully and make sure that the LED is the right way around. Note that if you put the LED in the wrong way, it will not do any harm; it just won't work.

Once you have the basic blink working, you can load the app from this project, which you can find with all the projects for this book in the public library called *PHOTON_BOOK*. The file is called *p_04_Morse_External_LED*.

There are just a few differences between this program and that of Project 3. They are all concerned with the change from using the built-in RGB LED and pin D7. The first change is the addition of a new variable to specify which pin to use:

```
int ledPin = D7;
```

The **setup** function is also slightly different, because now you need to set the **ledPin** to be an output:

```
void setup() {
    pinMode(ledPin, OUTPUT);
}
```

The only other code comes right at the end of the program, in the **flash** function:

```
void flash(int duration) {
    digitalWrite(ledPin, HIGH);
    delay(duration);
    digitalWrite(ledPin, LOW);
}
```

Now the **flash** function will turn the **ledPin** on and off rather than the built-in RGB LED.

Note that both the built-in tiny blue LED and the LED on the breadboard will flash the message. The internal LED on pin D7 consumes very little current and has negligible effect on the use of that pin as an output.

Attaching a Switch

Whereas an LED is probably the most common use of a digital output on a microcontroller like the Photon, a switch is probably the most common input.

Figure 4-5 shows the schematic diagram for attaching a switch to digital pin D3 of a Photon.

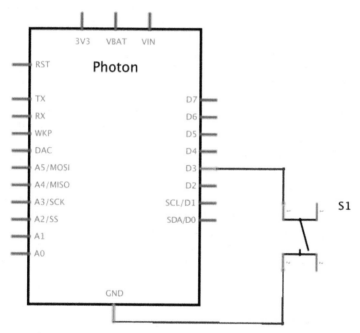

Figure 4-5. *Connecting a switch to a Photon*

The symbol for the switch is a little strange. You can see the switch contacts in the center of the switch and the two leftmost terminals of the switch are used. One terminal goes to the pin that will act as a digital input and one to GND. So when the switch is pressed, the digital input D3 will be connected to GND (0V).

The type of switch shown here, and used in the next project, is called a *tactile push switch* and usually has four pins. This gives it greater mechanical strength when it's soldered onto a PCB (or used with a breadboard). The two unused connections of the switch on the right just represent these extra pins.

In some schematic diagrams for using switches with a micro-controller, you may find a *pull-up resistor* connected between the digital input and 3.3V. The point of this resistor is to pull the input up to 3.3V (HIGH) until the switch is pressed, and then the switch will override the pulling effect of the resistor and make the input LOW. This is not necessary with the Photon or Core, as these devices' digital pins have built-in pull-up resistors that can be enabled in your code.

Switches and Outputs

The switch connects D3 to GND, which is just fine as long as D3 is set to be an input. However, if the program currently running on the Photon/Core uses D3 as a digital output and that output happened to be HIGH (3.3V), then pressing the switch would connect the 3.3V output to GND.

This is very bad and would quite likely damage the Photon or Core. So when using a Photon with digital inputs, it is a good idea to load the app first, before you start wiring up the hardware, so that there is no danger of a digital output being short-circuited to GND.

Digital Inputs

Before you can use a pin of the Photon/Core, you have to use the pinMode function to set its mode to be an input. Specifying that a pin should be used as an input indicates three possible types of digital input:

INPUT

> No pull-up resistors. This would typically be used for connecting the input to digital outputs of a chip or module.

INPUT_PULLUP

> Pull-up resistor enabled. Use this when connecting the digital input to a switch that will switch the input to GND.

INPUT_PULLDOWN

> Pull-down resistor enabled. Use this when connecting the digital input to a switch that will switch the input to 3.3V.

It might seem logical to use the last option and connect your switches to 3.3V so that pressing the button takes the input HIGH. However, it is more common to use the INPUT_PULLUP option and connect the switch to GND. This is partly because the Arduino boards do not have the third option, but also because if a pin is accidentally left as an output (see warning box), it will by default be LOW, reducing the chance of an accidental short circuit as a result of pressing the button.

To use the input connected to a switch, as I recommend and as shown in Figure 4-5, you would add the following command to your **setup** function:

```
pinMode(switchPin, INPUT_PULLUP);
```

In Project 5, the Morse Flasher project will be modified to start flashing only when a button is pressed.

Project 5. Morse Flasher with Switch

In this project, Project 4 will be expanded to include a tactile push switch. When the switch is pressed, the message will be flashed out in Morse code.

Parts

To build this project, you need the parts listed in Table 4-2 in addition to your Photon or Core.

Table 4-2. *Project 5 parts bin*

Part	Description	Appendix A code
LED1	Red, or for that matter, any other color LED	C2
R1	220Ω resistor	C1
S1	Tactile push switch	C3
	Male-to-male jumper wires	H4
	Half-sized breadboard	H5

These components, apart from the switch, are the same as for Project 4.

Software

You can find the software for this project in the file *p_05_Morse_Switch*. Get this loaded onto your Photon/Core before you wire up the switch.

The key difference between this program and the program for Project 4 is that there needs to be a new variable to specify the input pin:

```
int switchPin = D3;
```

In the setup function, this new pin needs to be specified as an input with pull-up resistor enabled:

```
void setup() {
    pinMode(ledPin, OUTPUT);
    pinMode(switchPin, INPUT_PULLUP);
}
```

Finally, the loop function needs to be changed so that the message is flashed only if the button is pressed. We can also do away with the delay at the end of loop, because the message will be flashed only when the button is pressed, rather than every time around the loop.

```
void loop() {
    if (digitalRead(switchPin) == LOW) {
        flashMessage(message);
    }
}
```

The if command tests to see if the result of doing a digital read on the switch pin is LOW. Remember that the input will be LOW when the switch is pressed; otherwise, it will be "pulled-up" HIGH.

Hardware

The breadboard layout for this project starts the same as that of Project 4. So, if you still have Project 4 on the breadboard, you can just add the switch and two jumper wires that lead to it (Figure 4-6).

Note that the push switch will just fit across the central divide between the two banks of breadboard rows. This will also ensure that it is the right way around and that the connections will go to a pair of pins that are separated by just one breadboard hole.

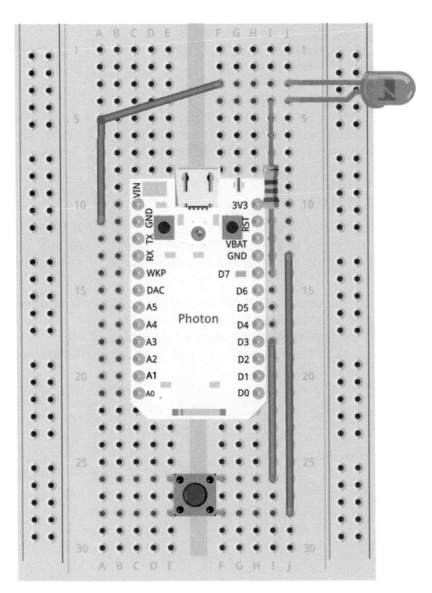

Figure 4-6. *Breadboard layout for Project 5*

Running the Project

Since you have already connected all the wiring, you should find that pressing the switch will start the Morse code flashing.

Analog Outputs

Although you have experimented with digital outputs, that is, turning something on and off, you may also want to control the brightness of an LED or the speed of a motor. To do this, you need to use an analog output.

Certain of the Photon's pins (pins D0, D1, D2, D3, A4, A5, WKP, RX, TX) and the Core's pins (D0, D1, A0, A1, A4, A5, A6, A7) can be used as analog outputs. These outputs are not true analog outputs, but are outputs that control the power to things that they are connected to by rapidly switching on and off (see the following sidebar about pulse-width modulation).

Pulse-Width Modulation

A pulse-width modulation (PWM) output of a Photon or Core will produce 500 pulses every second. Figure 4-7 shows what these pulses might look like.

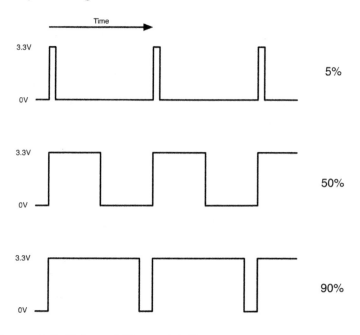

Figure 4-7. *Pulse-width modulation*

If the pulse is quite short (HIGH for, say, only 5% of the time), then a relatively small amount of energy flows to whatever the pin is powering. So if it is an LED, the LED will appear dim, because it is on for only 5% of the time. It switches so quickly that the human eye cannot tell that it's actually flashing really fast and just registers it as being dim. The same principle can be used to control a motor, only a small "kick" of energy is being given to the motor with each pulse.

If the PWM output is high for half of the time, then whatever is being powered (the load) will receive half-power; and if the PWM output is high for most of the time, then it will get a lot more power.

analogWrite

To control a PWM output, the `analogWrite` command is used. This command takes two parameters, as illustrated in this example:

```
analogWrite(ledPin, 127);
```

The first parameter is the pin to be used as a PWM output, and the second parameter is called the *duty cycle*. That is how long each pulse should be high. The value of the duty cycle is a number between 0, which means permanently off, and 255, which means permanently on. So a value of 127 will put the output at pretty much half power.

If you are using a pin as a PWM output, you still have to define it as an output by using `pinMode`.

An Example

As an example of using PWM, you can modify the hardware that you used in Project 5 to use a pin that is PWM-capable. To do this, move the resistor lead currently connected to D7 to A5. The revised breadboard layout is shown in Figure 4-8.

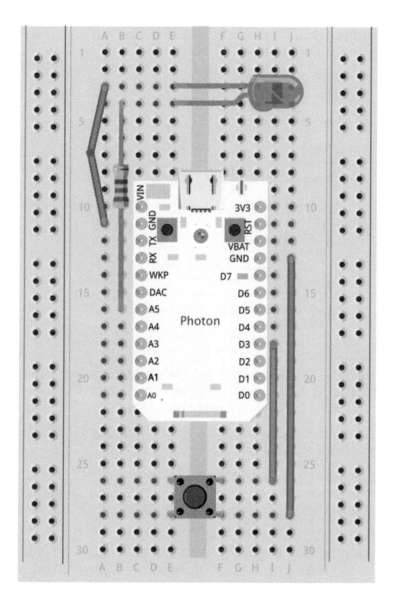

Figure 4-8. *Revised breadboard layout for PWM*

Flash the following program, which can be found in the *PHO-TON_BOOK* library as *ch_03_pwm_output*:

```
int ledPin = A5;
int switchPin = D3;

int brightness = 0;

void setup() {
    pinMode(A5, OUTPUT);
    pinMode(switchPin, INPUT_PULLUP);
}

void loop() {
    if (digitalRead(switchPin) == LOW) {
        brightness += 25;
        if (brightness > 255) {
            brightness = 0;
        }
        analogWrite(ledPin, brightness);
        delay(200);
    }
}
```

Initially, the LED will be off, but then each time that you press the button, it will get a little brighter. After a few button presses, the LED will cycle back around to being off.

A variable called **brightness** is used to keep track of the current duty cycle of the **ledPin**. This is initialized to 0.

The **loop** function starts with an **if** command that checks to see if the button is pressed. If it is, 25 is added to the variable **bright ness**. If the value of **brightness** exceeds 255, then it is set back to 0.

The duty cycle of the **ledPin** is set to the current value of **bright ness**, and then a delay of 200 milliseconds prevents the value from immediately changing again because the button is still pressed.

A Real Analog Output

The Photon has a "true" analog output that is not present on the older Core. This pin is immediately above A5 and is labelled DAC (Digital to Analog Converter). Unlike the pins that are capable of

PWM, the DAC pin does not use pulses but does actually change the voltage on its output between 0 and 3.3V.

To set the voltage of the DAC pin, you use `analogWrite`, just as you would with a PWM pin, for example:

```
analogWrite(DAC, 127);
```

Summary

In this chapter, you have learned the basics of digital inputs and outputs. As yet, we have not touched on analog outputs and inputs, but you will meet these in later chapters as the book progresses.

In Chapter 5, you will learn about the Internet of Things and how to control electronics over the Internet.

5/The Internet of Things

Aside from actually programming the Photon/Core over the Internet, everything we have done up to now could have been done with a device such as an Arduino without any Internet connection. While the Photon/Core is pretty good in the role of an Arduino replacement, to not use its Internet of Things capabilities is to miss out on its most exciting feature.

In this chapter, you will learn how to communicate with a Photon or Core over the Internet to issue commands to the device and to take readings from its sensors.

Functions

Many IoT devices require you to do some pretty hairy network programming to have your device communicate with the Internet. Not so the Photon and Core. Particle has managed to provide a simple and easy-to-use framework for using your Photon/Core over the Internet.

Later in this chapter, you will learn about reading information from the Photon/Core. In this section, I will concentrate on sending instructions to the Photon over the Internet. For example, in the next project, you will be able to send an HTTP post to a URL that then turns on an LED on the Photon.

The key to telling your Photon or Core to do things is the concept of a Function. To distinguish Particle Functions from C language functions in your programs, I will use an uppercase *F* when referring to Particle Functions.

A *Function* associates a command that you can send to the Photon/Core with a C function on that device's app. Whenever the device receives that command, it runs the associated C function. Rather like a regular C function, the function in the app also receives a parameter. The parameter is always of type String, and it's up to your function in the program as to how it uses that parameter, if indeed it uses the parameter at all.

 Actions and Functions
Note that in earlier versions of the software, Functions used to be called *actions*. This change was made to tie in better with the naming conventions of IFTTT (see Chapter 6).

The Photon has no practical limit on the number of Functions you can define, but the Core has a limit of four. This may sound restrictive, but remember that you can pass data to the function in its parameter and therefore do different things depending on that data.

In Project 6, you will make a simple example that will allow you to turn the built-in LED on pin D7 of the Photon/Core on and off from a command line. But first, if you are a Windows user, you will need to set up your environment.

Sending actions to a Photon/Core requires you to send HTTP post requests to the Particle service. A handy way to do this is using the command-line tool called *curl*. If you are a Mac or Linux user, the good news is that curl will already be installed and ready to use. If you are running Windows, you will need to install curl as described next.

Open your web browser to *http://curl.haxx.se/download.html* and scroll down to find the Win section of the downloads. Unless you have a really old computer, you will probably need the *Win64 - Generic* download.

Download the ZIP file and extract the single file it contains (*curl.exe*) to somewhere convenient, such as the desktop. Now start a command prompt by clicking the Start menu followed by the Run option and enter **cmd**.

This will open a command window like the one shown in Figure 5-1. Change the directory to wherever you saved *curl.exe* (in this case *Desktop*) and then run the command **curl** just to check that it's ready to use.

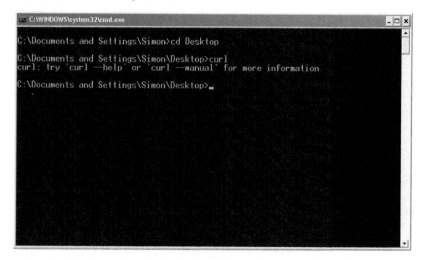

Figure 5-1. *Running curl on Windows*

The command **curl** is used to send HTTP requests without having to use a browser. You will use this tool to send commands over the Internet to your Photon/Core in the next project.

Project 6. Control an LED over the Internet

This project turns the LED attached to pin D7 of your Photon/Core on or off in response to commands over the Internet. Although in this case it's just an LED, the basic principle is pretty powerful; with the right hardware, you could be turning anything on or off.

The program for this app is called *p_06_LED_function*. The easiest way to get this onto your Photon/Core is to click the Libraries button in the Web IDE and then find the library called *PHOTON_BOOK*. You will find all the project programs used in this book as examples.

Select the file *p_06_LED_action* and then click the button USE THIS EXAMPLE. A copy of the program will open up in the editor, and you can flash it to your Photon/Core. Note that this will be your own copy of the original program, so you can modify it if you wish.

Software

The code for *p_06_LED_Function* is as follows:

```
int ledPin = D7;

void setup() {
    pinMode(ledPin, OUTPUT);
    Spark.function("led", ledSwitcher);
}

void loop() {
}

int ledSwitcher(String command) {
    if (command.equalsIgnoreCase("on")) {
        digitalWrite(ledPin, HIGH);
        return 1;
    }
    else if (command.equalsIgnoreCase("off")) {
        digitalWrite(ledPin, LOW);
        return 1;
    }
    return -1;
}
```

The code first defines a variable ledPin for the pin D7. This is the pin on the Photon/Core that has a tiny blue LED next to it. The setup function then defines this ledPin to be an OUTPUT.

The second line of setup is interesting:

```
Spark.function("led", ledSwitcher);
```

This is the line that defines the Function, giving it the name led. The second parameter, ledSwitcher, is the name of the C function in the program to be run when the led Function is invoked over the Internet.

The loop function is completely empty. You could do something in here if you wanted (for example, make a different LED blink), but it is perfectly OK to leave it empty. This program will do something only when the Function command is received.

When the `ledSwitcher` function is called as a result of a message arriving, it receives a `String` as its parameter. In the HTTP request that we are going to send, we will make sure that we either send **on** to turn the LED on or **off** to turn it off. Therefore, the `ledSwitcher` function needs to check the value of the command parameter and perform the appropriate digital write to the pin connected to the LED. If the command is either **on** or **off** (or for that matter, versions of those words in any letter case), the function will then return 1, indicating success. Any other value of command will result in the value -1 being returned by the function.

The return value will actually be returned as part of the response to the HTTP request, and this provides valuable information about whether the Function worked.

Security

Obviously, it would be potentially dangerous to allow every Particle user to have access to everyone's Photons. For this reason, there are two long tokens that help secure your Photon.

The first token is a device ID for the Photon or Core. Each of your devices will have a different device ID. To find the device ID for your device, click the Devices button on the IDE and select the device whose ID you want. The ID of that Photon/Core will then be displayed. Note that my devices are called A, B, C, and D, as shown in Figure 5-2.

To invoke a Function on that Photon/Core, you need to know this ID. A good place to keep this temporarily is as a comment after the last line of your program. For example:

```
// Device ID=55ff74062678501139071667
```

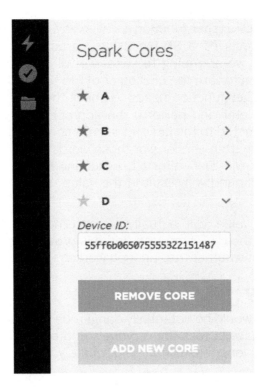

Figure 5-2. *Finding your device ID*

The second token is not specific to a Photon/Core, but rather to your whole Particle account. You will find this if you click the Settings button on the Web IDE, as shown in Figure 5-3.

You can at any time generate a new access token by clicking the Reset Token button.

You will also need this token when sending web requests, so you can paste another comment line with it into your program:

```
// Device ID=55ff74062678501139071667
// Access Token=cb8b348000e9d0ea9e354990bbd39ccbfb57b30e
```

Figure 5-3. *Finding your access token*

Trying It Out

Now that you have the program running on the Photon/Core, it will be sitting there waiting for a `led` Function to arrive. Let's give it what it wants by sending an HTTP request to the Particle web service. To do this, we will need to use the two tokens that we just discovered.

If you fetched *p_06_LED_Function* from the *PHOTON_BOOK* examples library, you will see these lines at the end of the program:

```
// To test with curl
curl https://api.spark.io/v1/devices/<deviceid>/led
-d access_token=<accesstoken> -d params=on
```

These lines tell us how to use `curl` to send an HTTP request to control the LED. But first you need to substitute your device ID and access token into the `curl` command. Once you have changed the comment command to include your tokens, it will look something like this:

```
// curl https://api.spark.io/v1/devices/
55ff74062678501139071667/led
   -d access_token=cb8b348000e9d0ea9e354990bbd39ccbfb57b30e
   -d params=on
```

Note that I have split this over multiple lines for readability, but the command that you paste into your command line must all be on one line.

Paste the `curl` command (don't include the `//` comment characters at the start) onto your command line, as shown in Figure 5-4.

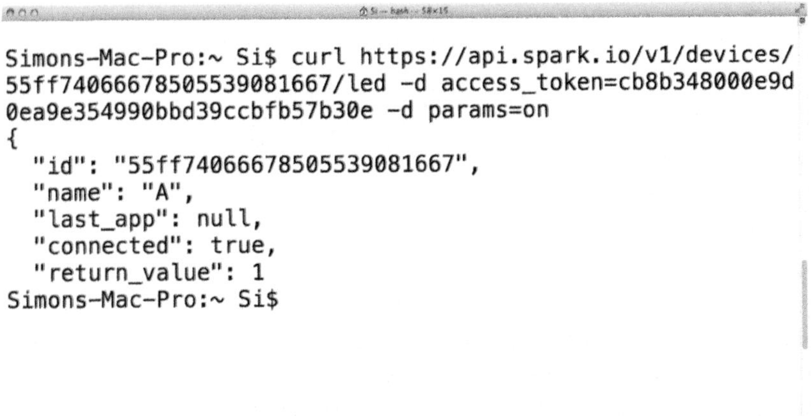

```
Simons-Mac-Pro:~ Si$ curl https://api.spark.io/v1/devices/
55ff74066678505539081667/led -d access_token=cb8b348000e9d
0ea9e354990bbd39ccbfb57b30e -d params=on
{
   "id": "55ff74066678505539081667",
   "name": "A",
   "last_app": null,
   "connected": true,
   "return_value": 1
Simons-Mac-Pro:~ Si$
```

Figure 5-4. *Controlling an LED with curl*

If all is well, the tiny blue LED on your Photon/Core should light up almost immediately, and you will see a response from your device containing the information that the `return_value` is 1.

To turn the LED off again, issue the same command (use the up arrow on your keyboard to recall the last command) and change on to off, and then press Enter to run the modified command.

You'll probably want to repeat this a few times, because it's pretty cool.

Interacting with loop

Now that Project 6 is working in its most basic form, let's make a small change to the program, so that instead of just turning the LED on and off, the `led` Function will set the LED blinking or stop it blinking. You can find the app code for this in the book examples with the name *p_06_LED_FUNCTION_BLINK*:

```
int ledPin = D7;
boolean blinking = false;
void setup() {
    pinMode(ledPin, OUTPUT);
    Spark.function("led", ledSwitcher);
}

void loop() {
    if (blinking) {
        digitalWrite(ledPin, HIGH);
        delay(200);
        digitalWrite(ledPin, LOW);
        delay(200);
    }
}

int ledSwitcher(String command) {
    if (command.equalsIgnoreCase("on")) {
        blinking = true;
        return 1;
    }
    else if (command.equalsIgnoreCase("off")) {
        blinking = false;
        return 1;
    }
    return -1;
}
```

This program has some code in the `loop` function. The code uses the *flag* variable called **blinking**. If **blinking** is set to **true**, the the LED will blink once. This will keep happening as long as **blinking** is **true**.

Now, the **ledSwitcher** function does not directly turn the LED on or off, but instead changes the value of the **blinking** flag.

Flash the program onto your Photon/Core and try issuing the same **curl** commands as you did for the previous program. Now the commands will start and stop the LED blinking rather than simply turning it on or off.

Running Functions from a Web Page

Although `curl` is good for testing Functions, it's in no way a slick user experience. What would be much nicer would be a web page that has buttons on it, so that when we click a button, it controls the LED (see Figure 5-5).

Figure 5-5. *Controlling an LED from a web page*

There is a large JavaScript library from Particle that encompasses everything from logging in and claiming devices through to calling Functions. You can learn more about this from the Particle documentation (*http://docs.particle.io/javascript/*), but for now, to keep things simple, let's use the more familiar jQuery JavaScript library on a web page to send HTTP posts in much the same way as when we were using `curl`, but with a more pleasant user interface.

Start by reflashing *p_06_LED_function* onto your Photon/Core. This is the program that simply responds to the `led` Function by turning the tiny blue LED on D7 on or off, depending on the value of its parameter.

Although eventually you might want to host the controlling page on a web server somewhere, you can test out the program with an HTML file on your computer and your browser. The HTML page for this can be found in the *html* section of the downloads for this book available from the book's GitHub repository (*https://github.com/simonmonk/photon_book*).

Download the file from GitHub onto your own computer. If you are familiar with GitHub, you may just want to clone the repository onto your computer to access all the example code in one go. If you do not have Git installed on your computer and don't want it, you can just download a ZIP archive of all the example code by going to the preceding URL and then clicking the Download ZIP button at the bottom right of the web page.

However you did it, find the *ch_05_led_control.html* file and open it in a text editor. You also need to replace the values near the top for `accessToken` and `deviceID` with your own access token and device ID.

Save the changes to the file and then open it in your browser by double-clicking it. It should look just like Figure 5-5.

When you click the ON button, it should light the tiny blue LED. Clicking OFF, not surprisingly, should turn the LED off.

Remember, the code running on the Photon/Core has not changed. All we have done here is use an HTML page to send the HTTP post when we press a button rather than use `curl`.

The HTML for the page is listed here:

```
<html><head>
<script src="http://ajax.googleapis.com/ajax/libs/jquery/1.3.2/
jquery.min.js" type="text/javascript" charset="utf-8"></script>
<script>
var accessToken = "cb8b348000e9d0ea9e354990bbd39ccbfb57b30e";
var deviceID = "55ff74066678505539081667";
var url = "https://api.spark.io/v1/devices/" + deviceID
          + "/led";

function switchOn(){
    $.post(url, {params: "on", access_token: accessToken });
}
function switchOff(){
    $.post(url, {params: "off", access_token: accessToken });
}
</script></head>

<body>
<h1>On/Off Control</h1>
<input type="button" onClick="switchOn()" value="ON"/>
<input type="button" onClick="switchOff()" value="OFF"/>
</body></html>
```

If you have never seen HTML before, take a look at an HTML primer such as *www.w3schools.com/html/html_intro.asp*.

This HTML file contains a mixture of HTML itself, including text that will appear on the page such as the heading and button labels. It also contains some code written in the JavaScript programming language.

JavaScript programs are often embedded in a web page to give it some intelligence or fancy user interface effects. In this case, the JavaScript is going to send the web requests to the Particle web service and hence to your Photon/Core.

If you have done any JavaScript programming in a web setting, you are likely to have come across the jQuery JavaScript library. This library includes loads of useful features for manipulating web pages. It also has a feature that allows it to make HTTP posts, which is just what you need to talk to your Photon/Core.

The jQuery library is imported from one of its homes on the Internet, and then we come to a second section of JavaScript, where some variables are defined for the access token and device ID. You should have already changed these values. The third variable, **url**, is constructed by splicing the value of **devi ceID** into the middle of the URL.

Next there are two nearly identical JavaScript functions. As you can see, JavaScript uses the same curly braces for blocks of code and has functions like C but with a different syntax.

If you look at the **switchOn** function, you can see that it contains the following:

```
$.post(url, {params: "on", access_token: accessToken });
```

The $ symbol is how you access the jQuery library functions, one of which is called **post**. It takes two parameters: **url** and what looks like another block of code. This is a JavaScript object and is a way of grouping together two values associated with names. The first value of **on** has the name **params**, and the second links the value of the **accessToken** variable with the name **access_token**.

The names **params** and **access_token** will become post data to the HTTP **post** command. These are equivalent to the **-d** options used in **curl**.

The body section of the HTML file is responsible for displaying the heading and buttons on the web page. Each of the input buttons has an onClick event. When you click one of the buttons, it runs the JavaScript function switchOn or switchOff depending on which button you click.

Security

Now that you can create a web page on your local hard drive to control your device, the temptation is to place the HTML file onto a web server, so that it can be accessed from anywhere.

You can do this, but not without compromising your security. Anyone doing View Source would immediately get access to your device ID and access token. They're there in the web page for all to see.

If this matters to you, you should host the page behind a login screen, or on a web server that runs within your network.

Project 7. Control Relays from a Web Page

In this project, you will build on what you have learned about Functions, using web pages to control a relay. The software for the project is very similar to the LED example that we have just been through. There is just a bit of renaming of code, to make it more specific to relays rather than LEDs, and the code is extended so that all four relays on the relay shield can be controlled.

The real difference in the project is that it uses a Photon/Core attached to a relay shield. This will allow us to switch all sorts of things on and off.

Figure 5-6 shows a Core plugged on top of a relay shield that is being used to turn any of the four relays on the module on or off. Note that one of the relays has an old-fashioned 12V electric bell attached to it. Figure 5-7 shows a simple web interface for controlling the relays.

AC Can Kill You

Although the relays on the relay board are technically capable of switching 120V AC at 10A, you should not use the board to control AC devices unless you are qualified to do so.

Some of the screw terminals on the board will be *hot* (also called *live*) if connected to AC, so anything made using this module that does control AC must be properly boxed and earthed so that there is no chance of any exposed metal being touched by unwary fingers.

Be warned: every year hundreds of people in the US alone die accidentally from 120V AC electrocution, not to mention the many more people who receive serious burns.

You can still have lots of fun with the relay module, but I strongly suggest that you stick to using it to switch low-voltage DC devices.

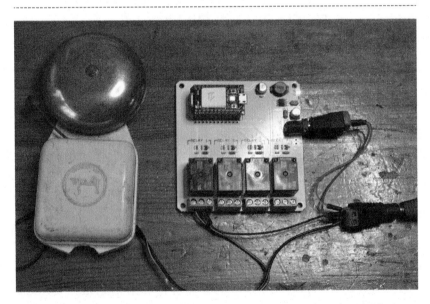

Figure 5-6. *Photon relay and electric bell—good and noisy!*

Relay Control

Figure 5-7. *A web page for controlling relays*

Parts

To build this project, you need the parts listed in Table 5-1 in addition to your Photon/Core.

Table 5-1. *Project 7 parts bin*

Part	Description	Appendix A code
Relay shield	Particle shield to control 4 relays	M1
12V power supply	12V 1A power supply	Q1
J1	DC jack-to-screw adapter (female)	H1
J2	DC jack-to-screw adapter (male)	H2
Bell	12V DC electric bell	Q2
Wire	Short lengths of multicore wire	H3

These days, most doorbells are of the electronic variety rather than the good old-fashioned bell complete with a solenoid-driven sounder. You should still be able to find such things in your local hardware store, though. Look for something on the packaging that tells you the current that the bell is likely to consume and make sure that the power supply you get will be sufficient to provide this current plus 0.2A for the Photon/Core and relay shield. Most bells should be fine. The extremely noisy bell I used consumed much less than 1A.

Many intruder alarm sirens and strobe lights also operate from 12V, so this may be an interesting alternative to the electric bell.

You can use male-to-male jumper wires for some of the connections to the bell and relay shield, but these tend to have low current-handling capabilities; so some multicore hookup wire or twin bell wire pulled apart into two strands is better. If you get into making electronics, you will find yourself scavenging odd bits of wire to keep for situations like this. You can also buy hookup wire in various colors.

Relays

A relay (Figure 5-8) is an electromechanical switch. It's very old technology, but reliable and easy to use.

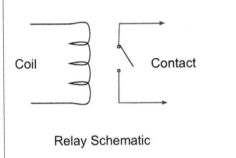

Relay Schematic

A Relay

Figure 5-8. *Relays*

A *relay* is basically an electromagnet that closes switch contacts. The fact that the coil of the electromagnet and the contacts are electrically isolated from one another makes relays great for things like switching AC-powered devices on and off from digital outputs like those of the Photon/Core. The coil of a relay takes just too much current to be driven directly from the digital output of a Photon, so the relay shield also has a small transistor to switch the relay coil. You will learn more about transistors in Chapter 8.

Whereas the coil of a relay is often energized by between 5V and 12V, the switch contacts can control high-power, high-voltage loads. For example, the relay photographed in Figure 5-8 claims a maximum current of 10A at 120V AC, as well as up to 10A at 24V DC.

Relays often have three switch terminals: COM (common), N.O. (normally open), N.C. (Normally Closed). When the relay coil is activated, the COM and N.O. terminals will be connected to each other. When there is no power to the relay coil, the N.C. and COM terminals will be connected. So, generally, you will want to use the N.O. and COM connections to switch whatever you are controlling, so that when the power to the Photon/Core is off, the device being switched will also be off.

Design

Figure 5-9 shows how you might wire up a 12V electric bell. You could also substitute 12V DC lighting or any other kind of 12V equipment for the bell. Just make sure that if the thing you are connecting has a positive and negative connection, the positive connection goes to the power supply and the negative connection to the relay. In this case, the bell does not have a positive and negative.

As you can see, the 12V power supply provides power to both the relay board and one side of the relay contacts that will switch the electric bell.

When the relay activates, the COM and N.O. (Normally Open) contacts of the relay will be connected together, completing the circuit from the bell to the 12V power supply, making the bell sound.

Construction

This is an easy project to make: no soldering is required, and everything is hooked up using screw terminals. Use Figure 5-9 as a reference while you wire things up.

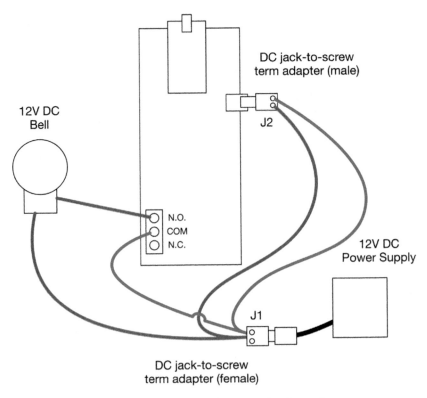

Figure 5-9. *Wiring an electric bell to a relay board*

Step 1. Test and Wire the Bell

If the bell does not have flying leads, you will have to attach leads to it. This will depend on your bell, but it may well have screw terminals too. Having attached wires to it, it is not a bad idea to connect the two wires directly to the power supply, just to check that you have the right connections and that the power supply and bell are compatible.

Step 2. Complete the Wiring

The remainder of the wiring centers around the J1 adapter. The tricky bit here is that each of the screw connectors on the adapter must hold two wires. The positive connection on J1 must have one wire going off to the positive connection of J2, and another wire to the bell. The negative terminal of J1 must contain one lead to the negative connection on J1, and another wire to the COM screw terminal of relay 1.

Be very careful to make sure that the positive (+) connections of J1 and J2 are connected together.

Step 3. Plug in the Power Supply

Plug the DC jack from the power supply into J1 and switch it on. If you have a smartphone, you can use the Tinker app to turn pin D0 on and off, but you will need to factory reset your Photon for the Tinker app to be reinstalled. This should turn the bell on and off. Pin D0 is connected to relay 1 on the relay shield.

Now you are ready for the software!

Software

The software for this project is in two parts: the app to be run on the Photon/Core, and the HTML web page containing Java-Script commands that will communicate with the Photon/Core.

Photon Software

The code for the Photon/Core is remarkably compact (*p_07_LED_Relays*). The full listing is given next. You can find this code along with all the code used in the book. See the introduction to Project 6 for instructions on getting hold of this.

```
int relayPins[] = {D0, D1, D2, D3};
void setup() {
    for (int i = 0; i < 4; i++) {
        pinMode(relayPins[i], OUTPUT);
    }
    Spark.function("relay", relaySwitcher);
}

void loop() {
}

int relaySwitcher(String command) {
    if (command.length() != 2) return -1;
    int relayNumber = command.charAt(0) - '0';
    int state = command.charAt(1) - '0';
    digitalWrite(relayPins[relayNumber-1], state);
```

```
        return 1;
    }
```

The first interesting thing about this program is that it uses an array to access the four digital pins that control the relays. The pins that control the relays are D0 to D3, where D0 controls relay 1, D1 relay 2, and so on.

The **setup** function uses a loop to initialize all four pins as outputs and then defines a Function called **relay** that will invoke the C function **relaySwitcher**.

The **relaySwitcher** function is able to switch any of the four relays on and off, not just the one we have connected to a bell. To do this, it uses its **command** parameter. The **relaySwitcher** function expects the **command** parameter to be exactly two characters long, so the first thing the function does is to check that this is the case. If **command** isn't two characters long, -1 is returned to indicate an error.

Assuming that the **command** parameter is two characters, the first character will be the number of the relay to control (1 to 4), and the second character will be 1 to turn the relay on or 0 to turn it off.

The code uses a useful trick to convert the character value of the relay number, which is of type **char**, to an **int**:

```
int relayNumber = command.charAt(0) - '0';
```

The trick is to subtract the character value of 0 from the character value of the first character in the command (**command.charAt(0)**).

The **digitalWrite** command uses the array and the relay number to decide which pin to set to either 0 or 1 from the value passed in the second character of the command.

Web Page Software

When it comes to the web page software to control the relays, we will start gently, with a modified version of the HTML and JavaScript that we used in "Running Functions from a Web Page" on page 88. The problem with that solution is that if something goes wrong—perhaps the Photon/Core isn't connected to the network—there is no way of knowing from the web page whether the command was successful and the relay actually turned on.

To get around this problem, there will be a second, more complicated version of the web page that uses the response from the device to provide feedback on the success or failure of the Function command.

Both versions are available from the book's GitHub page (*https://github.com/simonmonk/photon_book*) in the *html* section of the repository. The first version (listed next) is called *ch_06_relays.html,* and the improved version is called *ch_06_relays_fb.html.*

```html
<html><head>
<script src="http://ajax.googleapis.com/ajax/libs/jquery/1.3.2/
jquery.min.js" type="text/javascript" charset="utf-8"></script>
<script>
var accessToken = "cb8b348000e9d0ea9e354990bbd39ccbfb57b30e";
var deviceID = "54ff72066672524860351167";

var url = "https://api.spark.io/v1/devices/" + deviceID
          + "/relay";
function setRelay(message){
    $.post(url, {params: message, access_token: accessToken });
}
</script>
</head>
<body>
<h1>Relay Control</h1>
<table>
<tr>
<td><input type="button" onClick="setRelay('11')"
                value="Relay 1 ON"/></td>
<td><input type="button" onClick="setRelay('10')"
                value="Relay 1 OFF"/></td>
</tr>
<tr>
<td><input type="button" onClick="setRelay('21')"
                value="Relay 2 ON"/></td>
<td><input type="button" onClick="setRelay('20')"
                value="Relay 2 OFF"/></td>
</tr>
<tr>
```

```
<td><input type="button" onClick="setRelay('31')"
            value="Relay 3 ON"/></td>
<td><input type="button" onClick="setRelay('30')"
            value="Relay 3 OFF"/></td>
</tr>
<tr>
<td><input type="button" onClick="setRelay('41')"
            value="Relay 4 ON"/></td>
<td><input type="button" onClick="setRelay('40')"
            value="Relay 4 OFF"/></td>
</tr></table>
</body></html>
```

The first part of the web page is exactly the same as the example from "Running Functions from a Web Page" on page 88. Variables are defined for the accessToken and deviceID that you will need to change to match your own access token and device ID.

The separate functions that turned the LED on and off are replaced by a single JavaScript function called setRelay. The setRelay function is passed as a parameter called message that will contain the two-character code for controlling the relay, the first character being the relay number and the second 1 or 0 to indicate on or off.

The buttons in the user interface call setRelay with the right value for the button being pressed.

After editing the deviceID and accessToken variables for your values, open the file in your browser by double-clicking it. The browser window should look like Figure 5-10.

Figure 5-10. *A basic web interface for relays*

When you click the Relay 1 ON button, the relay should click, the little light next to it will come on, and most important, the bell will sound. You'll probably want to turn it off again quickly, so click Relay 1 OFF.

Try the other buttons to make sure that you can control all the relays.

One snag with this interface to the relays is that if you turn off the power to the relay shield and hence the Photon/Core, the web page will carry on behaving exactly as it currently does, without any indication that nothing is actually happening at the other end of the connection.

This second version of the web page, which is the interface shown previously in Figure 5-7, has a single button for each relay. Clicking the button toggles the relay for that button on and off, and changes the color of the button to indicate that the relay is on. However, the button does not change color to indicate that the relay is on until it receives confirmation from the Photon/Core that the Function worked successfully.

The code for this web page (*ch_06_relays_fb.html*) is not listed in full next as it's some 60 lines long. You may wish to have the file open in an editor while the code is explained.

```
<style type="text/css">
    .on  { background:green; }
    .off { background:red; }

    input[type=button] {
      color: white;
      font-size: 2.5em;
    }
</style>
```

The **style** block defines the colors to be used with the button for its on and off states and also increases the size of the font on the buttons.

Before looking at the JavaScript function **toggleState**, where all the action happens, it is perhaps worth looking at how it is called from one of the buttons:

```
<input type="button" id="btn" value="Relay 1" class="off"
    onclick="toggleState(this, '1')" />
```

When a button is clicked, `toggleState` is called with two parameters. The first parameter is `this`, which refers to the button being pressed. The `toggleState` function will need this to determine the current state of the button. The second parameter is the relay number as a string.

The `toggleState` function is listed here:

```
function toggleState(item, relayNumber){
    function setRelay(message){
        $.post(url, {params: message, access_token:
            accessToken}, callback);
    }
    function callback(data, status){
        if (data.return_value == 1) {
            if (item.className == "on") {
                item.className="off";
            }
            else {
                item.className="on";
            }
        }
        else
        {
            alert("Could not Connect to Photon");
        }
    }
    if(item.className == "on") {
        setRelay(relayNumber + "0");
    } else {
        setRelay(relayNumber + "1");
    }
}
```

In this case, the functions **setRelay** and **callBack** are defined within **toggleState**, so that they can have access to **toggleS tate**'s variables, in particular to its **item** parameter. The **setRelay** function doesn't actually need this, but it makes sense to keep both functions together as they are closely related.

The **setRelay** function is very like the earlier version of this function in the simplified version of this page. There is one important difference: the call to **$.post** now includes an additional parameter of **callback**. This is a reference to the other function inside **toggleState**. The **callback** function will automatically be run when the web request to the Photon/Core completes. We will return to the `callback` function shortly, but first let's look at the main body of the `toggleState` function:

```
if(item.className == "on") {
    setRelay(relayNumber + "0");
} else {
    setRelay(relayNumber + "1");
}
```

The `item` variable contains a reference to the button that was clicked, and its `className` will either be **on** or **off**, depending on the current state of the button. The `setRelay` function is then called, appending 0 or 1 to the relay number as appropriate. Note that at this point the appearance of the button has not changed; this will happen only if the communication with the device is successful.

When the request to the Photon/Core completes, the `callback` function will be invoked. Its first parameter (**data**) will be the response from the Photon/Core via the Particle web service. This is the same response that you are used to seeing as a result of a `curl` command to invoke a Function. The `callback` function picks out the `return_value` from this response, and as long as it is 1, indicating success (-1 indicates failure), the button appearance is changed by setting its `className`.

If `callback` does not get a response of 1, it alerts with an error message saying, "Could not connect to Photon."

Remembering to change the `accessToken` and `deviceID` variables in your file, open a browser on the page and try turning some relays on and off.

Project 8. Morse Code Text Messages

Projects 5 and 6 both flashed out a Morse code message, but suffered from the limitation that the message was fixed. The only way to change the message was to alter the value of the message variable in the program and then flash it onto your Photon/Core.

In this project, the message to be sent as Morse code is sent as the parameter to a Function. The maximum length of this parameter is 64 characters, so you will need to keep your message below that number of characters.

The project also has a web page where you can specify the message to send (Figure 5-11).

Figure 5-11. *Sending Morse code from a web page*

For added fun, the project will also gain a buzzer so that the Morse code is both flashed and beeped.

Parts

To build this project, you need the parts listed in Table 5-2 in addition to your Photon/Core.

Table 5-2. *Project 8 parts bin*

Part	Description	Appendix A code
LED1	Red or any other color LED	C2
R1	220Ω resistor	C1
R2	1kΩ resistor	C5
S1	Piezo buzzer	C4
	Male-to-male jumper wires	H4
	Half-sized breadboard	H5

These components, apart from the buzzer, are the same as for Project 4.

Software

Most of the code here is our now standard Morse sending code. You can find the full program to flash onto your device under the filename *p_08_Morse_Function* in the *PHOTON_BOOK* library in the Web IDE.

The first difference between this and our earlier Morse programs is the addition of a new pin for the buzzer:

```
int buzzerPin = D1;
```

Because sending a Morse code message can take a while, it would be better if the handler for the Function to send a Morse message would return immediately with a status number of 1 or -1 rather than wait for the entire message to be sent. That way, the web page from which the message is sent could provide confirmation that the Morse code was sending on the Photon/Core. To allow this to happen, there needs to be a new **boolean** value that we can set to **true** to indicate that flashing should start:

```
boolean flashing = false;
```

The **setup** function now needs an extra line to initialize this pin as an output. The **setup** function also contains the line to define the Function that will be invoked whenever the Photon/Core receives a Function command to send a message. This will call the function **startFlashing**:

```
void setup() {
    pinMode(ledPin, OUTPUT);
    pinMode(buzzerPin, OUTPUT);
    Spark.function("morse", startFlashing);
}
```

The **loop** function will be responsible for doing the flashing, but only if the **flashing** variable is set to **true**:

```
void loop() {
    if (flashing) {
        flashMessage(message);
        flashing = false;
    }
}
```

The **startFlashing** function handles the Function called **flash**. The **startFlashing** function does not actually do the flashing. Instead, it first checks that the length is less than or equal to the maximum allowed parameter size of 64 characters. If it's not too long, the **flashing** variable is set to **true** so that the main loop can start flashing (and buzzing) the message. If all is good, 1 is returned; otherwise, -1 is returned to indicate an error:

```
int startFlashing(String param) {
    if (param.length() <= 63) {
        message = param;
        flashing = true;
        return 1;
    }
    else {
        return -1; // message too long
    }
}
```

The only other change to the code is in the **flash** function that must now buzz at the same time:

```
void flash(int duration) {
    digitalWrite(ledPin, HIGH);
    tone(buzzerPin, 1000);
    delay(duration);
    noTone(buzzerPin);
    digitalWrite(ledPin, LOW);
}
```

Buzzing is accomplished using the **tone** and **noTone** commands. These commands can be used only on the pins D0, D1, A0, A1, A4, A5, A6, A7, RX, and TX.

The function **tone** takes two parameters: the pin to play a tone on and the frequency of that tone in Hertz (cycles per second). Once started using **tone**, the pin will continue to oscillate until the **noTone** command, with a parameter of that pin, is called.

Flash the program onto your Photon or Core so that the app is installed before you start attaching hardware to it.

You saw the web page for controlling the Morse code sender in Figure 5-11. The code for this is listed next. This can be found with the book downloads in the *html* section of the repository. The file is called *p_08_morse_function.html*:

```
<html>
<head>
<script src="http://ajax.googleapis.com/ajax/libs/jquery/1.7.2/
jquery.min.js" type="text/javascript" charset="utf-8"></script>
```

```
<script>
var accessToken = "cb8b348000e9d0ea9e354990bbd39ccbfb57b30e";
var deviceID = "54ff72066672524860351167"
var url = "https://api.spark.io/v1/devices/" + deviceID
          + "/morse";

function callback(data, status){
    if (data.return_value == 1) {
        alert("Your Message is being sent");
    }
    else
    {
        alert("Could not Connect to Photon");
    }
}

function sendMorse(){
    message = $("#message").val();
    $.post(url, {params: message, access_token: accessToken},
                callback);
}
</script>
</head>

<body>
<h1>Morse Code Sender</h1>
<input id="message" value="Hello World" size="64"/>
<input type="button" value="Send" onclick="sendMorse()" />
<br/>
</body>
</html>
```

The web page code follows much the same pattern as the previous examples. The URL is constructed using the accessToken and deviceID variables that must be changed to match those of your account and device.

In this example, the callback function does not need to be nested within the sendMorse function, as it does not need access to any of its variables or parameters. The callback function pops up an alert dialog box to indicate success or failure.

The sendMorse function first extracts the message to be sent from the field on the web page with the ID of message and then supplies this as the parameter to the HTTP request that will trigger the morse Function on the Photon/Core.

Hardware

Figure 5-12 shows the breadboard layout for the project.

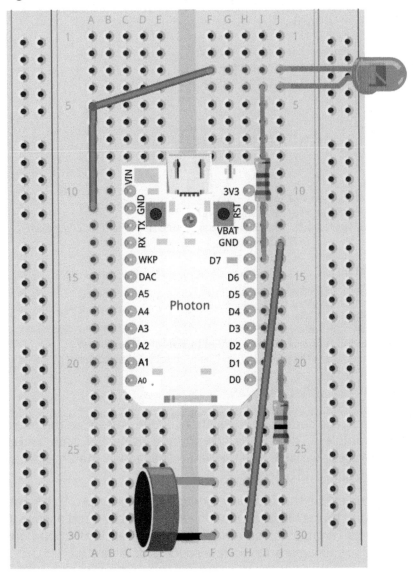

Figure 5-12. *Breadboard layout for Morse code with buzzer*

You can tell the 220Ω and 1kΩ resistors apart because the 220Ω resistor has stripes that are red, red, brown, and the 1kΩ resistor has brown, black, red. The buzzer can be placed either way around, but remember that the LED must have the longer positive lead on the same row as the resistor R1.

When the breadboard is fully assembled, power up the Photon/Core; it's time to try out the project.

Using the Project

Open the web page for this project in your browser, enter a message into the text field, and click the Send button. After a few moments, the Morse code message that you typed will start to be sounded by the buzzer while at the same time flashed by the LED. Remember that the maximum message length is 63 characters.

Variables

So far, all of our Internet of Things projects have been about telling the Photon/Core to do something using Functions. Being able to read information from the Photon/Core is just as important. Projects 9 and 10 are both concerned with using sensors with a Photon over the Internet.

To fetch information from a Photon/Core, you need to use Particle Variables. Defining variables in your app allow you to send requests to your Photon/Core that will bring back a value. Rather like Functions, Particle Variables can easily be confused with regular C variables, so I will use an uppercase *V* when referring to Variables.

Variables have other similarities with Functions. They have an identifier, which is a short string used to identify the Variable, so that you know what to ask for in a web request. They are also defined in the **setup** function, but rather than being associated with a C function, they are associated with a C variable, whose value they will report. It is the job of your app to keep updating the values in the C variable, so that when a request comes in for a Variable, the value returned is up to date.

In the next project, you will see how this works in a simple example, but first you need to learn a little about how you can connect sensors to a Photon/Core and take readings from them.

Analog Inputs

You will find six pins down the lefthand side of the Photon, labelled A0 to A5. The Core has two extra analog inputs, A6 and A7. Although these pins can be used as digital inputs or outputs just like D0 to D7, the *A* pins are capable of analog input.

Whereas a digital input can tell only if something is on or off, an analog input can measure the voltage at a pin as a number between 0 and 4095, where a reading of 0 indicates 0V and 4095 means 3.3V.

Maximum Voltage of 3.3V

Do not be tempted to attach an analog input to a voltage higher than 3.3V. This is likely to damage your Photon.

If a sensor whose output voltage varies somewhere in the range 0 to 3.3V, you can connect it to one of the analog pins and read the voltage and therefore work out a value for the property that the sensor is measuring.

As an example of this, load up the file *ch_05_analog_ins* from the *PHOTON_BOOK* example library. This is listed here:

```
int reading = 0;
int analogPin = A0;

void setup() {
    Spark.variable("analog", &reading, INT);
}

void loop() {
    reading = analogRead(analogPin);
}
```

Here the program starts with a C variable (**reading**). This is the C variable whose value will be returned when a web request

Function is sent to the Photon/Core. The other variable (`analog Pin`) just identifies the analog pin to use, in this case, A0.

The link between the C variable and the Particle Variable takes place in the **setup** function. The first parameter (`analog`) is the name given to the Variable, and the second parameter (`&read ing`) identifies the C variable that the analog Variable is to use. The & symbol is a C language feature that indicates the memory address of a variable. If the second parameter were just `reading` rather than `&reading`, then the value of `reading` would be substituted into that parameter, and, since `reading` is 0 at the time, the Variable values would never change. Using & allows the Photon/Core's firmware to defer fetching the value of the variable until it needs the value of the Variable.

Flash the program onto your Photon/Core. To test it, you can use a web browser because, unlike Functions, Variables use HTTP get requests, which you can issue simply by entering a URL into the address bar of your browser. So, open a browser window and enter the following URL into it:

```
https://api.spark.io/v1/devices/<DEVICE ID>/analog?
access_token=<ACCESS TOKEN>
```

Remember to substitute your device ID and access token into the URL. The result from the web page will look like Figure 5-13.

Figure 5-13. *Requesting a Variable by using the browser*

Refresh the page a few times and you should see the value of `result` change a little each time. This is not really surprising as the analog does not have anything connected to it; it is said to be *floating*.

Using a male-to-male jumper wire, connect A0 first to GND (Figure 5-14) and then to 3.3V (Figure 5-15), and each time refresh the web page. You should find that with A0 connected to GND, you will get a result of 0. With it connected to 3.3V, you will get something close to 4095.

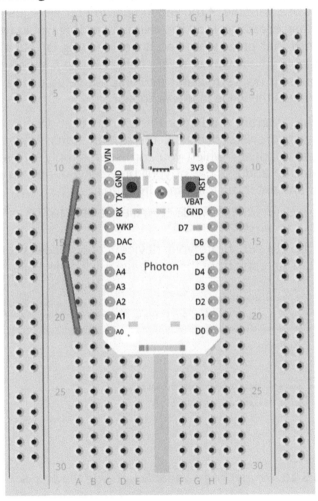

Figure 5-14. *Connecting A0 to GND*

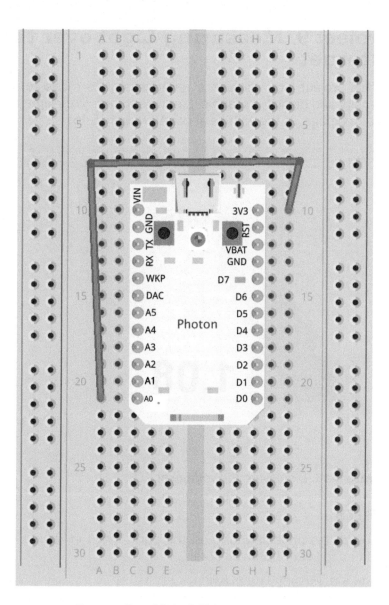

Figure 5-15. *Connecting A0 to 3.3V*

In Project 9, we will use a photoresistor as a light sensor, whose value can be retrieved using a Variable web request.

Project 9. Measuring Light over the Internet

In this project, you will use a photoresistor to measure light intensity and display a graphical representation of the light level on a web page, as the voltage at pin A0 (Figure 5-16).

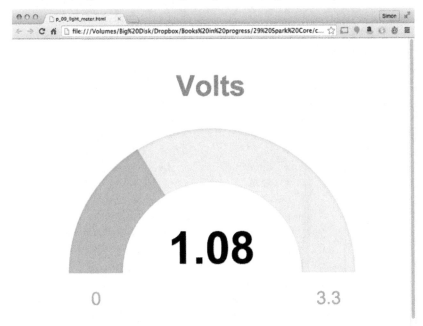

Figure 5-16. *Displaying the voltage at A0*

Parts

To build this project, you need the parts listed in Table 5-3 in addition to your Photon or Core.

Table 5-3. *Project 9 parts bin*

Part	Description	Appendix A code
R1	Photoresistor (1kΩ)	C6
R2	1kΩ resistor	C5
	Male-to-male jumper wires	H4
	Half-sized breadboard	H5

Software

The app for this project can be found in the file *p_09_Lightmeter* within the *PHOTON_BOOK* library in the Web IDE.

```
int reading = 0;
double volts = 0.0;
int analogPin = A0;

void setup() {
    Spark.variable("analog", &reading, INT);
    Spark.variable("volts", &volts, DOUBLE);
}

void loop() {
  reading = analogRead(analogPin);
  volts = reading * 3.3 / 4096.0;
}
```

This code is very similar to that of the example in the previous section. However, this version has two Variables, `reading` and `volts`. In fact, you can have up to 10 Variables in your app.

The `reading` Variable is associated with the `reading` C variable and returns just the same result as the previous example. However, the second Variable (`volts`) will return the actual voltage at the analog pin. To calculate this voltage, the raw reading is first multiplied by 3.3 and then divided by 4096 (the maximum range of `reading`).

The fancy graphical display is courtesy of a really handy JavaScript plug-in called JustGage (*http://justgage.com*). The web page for this can be found in the file *p_09_light_meter.html* along with the JustGage JavaScript libraries.

```
<html>
<head>
<script src="http://ajax.googleapis.com/ajax/libs/jquery/1.7.2/
jquery.min.js" type="text/javascript" charset="utf-8"></script>
<script src="raphael.2.1.0.min.js"></script>
<script src="justgage.1.0.1.min.js"></script>

<script>
var accessToken = "cb8b348000e9d0ea9e354990bbd39ccbfb57b30e";
var deviceID = "54ff72066672524860351167"
var url = "https://api.spark.io/v1/devices/" + deviceID + "/
```

```
volts";

function callback(data, status){
    if (status == "success") {
        volts = parseFloat(data.result);
        volts = volts.toFixed(2);
        g.refresh(volts);
        setTimeout(getReading, 1000);
    }
    else {
        alert("There was a problem");
    }
}

function getReading(){
    $.get(url, {access_token: accessToken}, callback);
}
</script>
</head>

<body>
<div id="gauge" class="200x160px"></div>

<script>
var g = new JustGage({
    id: "gauge",
    value: 0,
    min: 0,
    max: 3.3,
    title: "Volts"
});
getReading();
</script>

</body>
</html>
```

The JustGage library and the Raphael library that it uses are both imported and variable-defined for the **accessToken** and **deviceID**. Don't forget to change these values for your account and device.

Jumping now to the end of the file, the JavaScript function **getReading** is called. This constructs the HTTP request to be sent and attaches the **callback** function to it.

When the request responds, the `callback` function checks to see if the request was successful and, if it was, retrieves the voltage reading from the result, converting it from a string to a float by using the command `parseFloat`. Because the number may have a lot of digits after the decimal point, the number is truncated to two decimal places by using the `toFixed` function.

The gauge display is updated with the new voltage reading by calling `g.refresh`, after which another call to `getReading` is scheduled for one second later, using `setTimeout`.

The gauge control itself requires a `div` tag with an ID of `gauge`, where it should draw itself and a short section of script that initializes it. Here you can change the minimum and maximum values for the gauge as well as the title.

Hardware

Figure 5-17 shows the breadboard layout for Project 9. Both the resistor and photoresistor can be connected either way around. They form a voltage divider (see "Photoresistors and Voltage Dividers" on page 119) that changes the voltage at A0 depending on the quantity of light falling on the photoresistor.

Once you have connected the breadboard, power up your Photon/Core, and the project is ready to use.

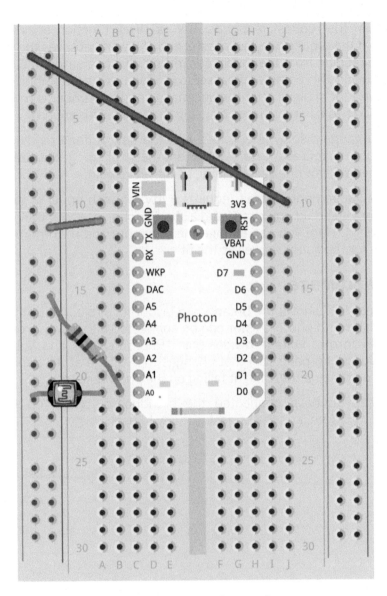

Figure 5-17. *Breadboard layout for Project 9*

Photoresistors and Voltage Dividers

In this project, a fixed-value resistor is used with a photoresistor whose resistance varies according to how much light falls on it. They are in an arrangement called a *voltage divider* (see Figure 5-18).

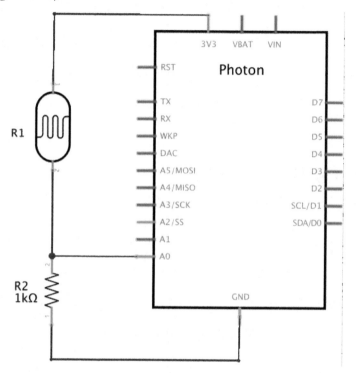

Figure 5-18. *Voltage dividers*

If the photoresistor had just the right amount of light falling on it to make its resistance the same as R2 (1kΩ), the 3.3V would be split equally across the two resistors and the voltage at A0 would be half of 3.3V (1.66V). The lower the resistance of the photoresistor (more light), the less voltage it will grab and therefore the more voltage there will be across R2 and the higher the voltage at A0.

Excessive Data Polling

Repeatedly reading data in this way is called *polling*. In this case, the web page is requesting a Variable value from a Photon/Core every second. If tens or hundreds of thousands of Photon users are all doing this kind of thing at the same time, the Particle service is going to be hammered pretty hard.

So bursts of polling once a second while you are using it are fine, but the socially responsible thing to do would be to make sure that you close the browser window when you are not using it, or set the polling rate to, say, once every 10 seconds if it's going to be running all the time.

Using the Project

Open the web page *p_09_light_meter.html*, and the window should look like Figure 5-16.

Cover the photoresistor with your hand and you should see the reading fall; hold it up to the light, and the meter should swing over to the right.

There are other types of resistive sensors that could be substituted for the photoresistor, including strain gauges, variable resistors (like volume controls), and even some gas detectors.

In the next project, you will use a different type of sensor to provide remote temperature monitoring.

Project 10. Measuring Temperature over the Internet

This project is very similar to the last, but rather than measure light intensity with a photoresistor, it uses a digital temperature sensor chip. The gauge display is perfect for this. Figure 5-19 shows the display in degrees F, but this can easily be changed to degrees C.

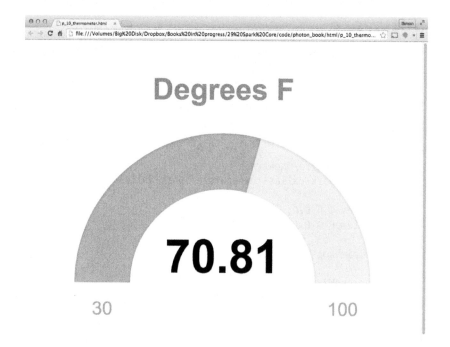

Figure 5-19. *An Internet thermometer*

Parts

To build this project, you need the parts listed in Table 5-4 in addition to your Photon/Core.

Table 5-4. *Project 10 parts bin*

Part	Description	Appendix A code
IC1	DS18B20 temperature sensor	C7
R1	4.7kΩ resistor	C9
C1	100nF (0.1uF) capacitor	C8
	Male-to-male jumper wires	H4
	Half-sized breadboard	H5

The temperature sensor used is actually a chip enclosed in a watertight container with a lead attached to it, so you can place the sensor some distance away from the Photon/Core if you want.

Software

The app for this project can be found in the file *p_10_Thermometer_Dallas* within the *PHOTON_BOOK* library in the Web IDE. The name *Dallas* is used, because this common type is made by Dallas Semiconductors.

The program includes two Variables: one that reports the temperature in degrees F and one in degrees C.

```
#include "OneWire/OneWire.h"
#include "spark-dallas-temperature/spark-dallas-temperature.h"

double tempC = 0.0;
double tempF = 0.0;

int tempSensorPin = D2;

OneWire oneWire(tempSensorPin);
DallasTemperature sensors(&oneWire);

void setup() {
    sensors.begin();
    Spark.variable("tempc", &tempC, DOUBLE);
    Spark.variable("tempf", &tempF, DOUBLE);
}

void loop() {
    sensors.requestTemperatures();
    tempC = sensors.getTempCByIndex(0);
    tempF = tempC * 9.0 / 5.0 + 32.0;
}
```

The first thing to note about this program is that it uses two libraries, indicated by the #include statements at the top. Libraries are a way of sharing code for anyone to use, and, because interfacing to the DS18B20 chip is a little tricky, Tom de Boer has converted an original Arduino library to work with the Photon and Core. This library (*spark-dallas-temperature*) relies on another library called OneWire to handle the serial communication with the temperature sensor chip.

For your app to work, as well as having the #include commands at the top, you also need to add the libraries to the app. To do this, first click the Libraries button in the Web IDE and then type

"Dallas" into the search box beneath Community Libraries (Figure 5-20).

Figure 5-20. *Using the DallasTemperature library in an app*

Click the INCLUDE IN APP button and then scroll down to *P_10_THERMOMETER_DALLAS*, select it, and then click ADD TO THIS APP. Repeat the same process, searching for the library OneWire.

Turning to the code itself, after the variable `tempSensorPin` has been defined as D2, the following definition is called to start the OneWire serial bus running on that pin:

```
OneWire oneWire(tempSensorPin);
```

The next line instructs the DallasTemperature library to use that interface when communicating to temperature sensors. You can connect a whole load of temperature sensors to the same pin if you want.

The `setup` function contains a call to `sensors.begin()` to start the sensors monitoring the temperature.

The `loop` function contains the code to actually get hold of a reading. This is a two-stage process. First, `sensors.requestTemperatures` is called, which gets readings from all the sensors (even though in this case there is only one). The second step is to access the reading itself using `getTempCByIndex`. The index supplied as a parameter is 0 for the first sensor attached to the Photon/Core. If there was a second DS18B20 attached, it would be index 1, and so on.

The temperature in degrees F is then calculated using a standard formula.

The web page is almost identical to the light-sensor page, but with a few changes to reflect that it is the temperature being displayed. The web page can be found in the file *p_10_thermometer.html*.

Hardware

The breadboard layout for the project is shown in Figure 5-21.

The temperature sensor has four connectors. One is a connection to the braided screening within the cable. This does not need to be connected. The other wires are red (positive supply), black (GND), and yellow (the data signal). The resistor must be connected between the data signal and positive supply, and the capacitor between the positive supply and GND.

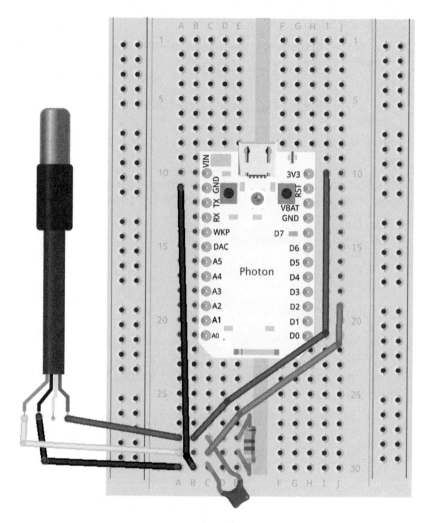

Figure 5-21. *Breadboard layout for Project 10*

Using the Project

To use the project, upload the app to the Photon/Core and open the web page *p_10_thermometer.html* in your browser.

If you want to change the temperature display to be in degrees C rather than F, open *p_10_thermometer.html* in an editor and make the following changes:

Change the lines

```
var url = "https://api.spark.io/v1/devices/" + deviceID + "/
tempf";
```

to read

```
var url = "https://api.spark.io/v1/devices/" + deviceID + "/
tempc";
```

In other words, use the `tempc` Variable in place of the `tempf` Variable.

You also need to change the title of the gauge and its range, so modify the script block at the end of the file to look like this:

```
<script>
var g = new JustGage({
    id: "gauge",
    value: 0,
    min: 0,
    max: 30,
    title: "Degrees C"
});
getReading();
</script>
```

Reload the page in your browser, and it should now look like Figure 5-22.

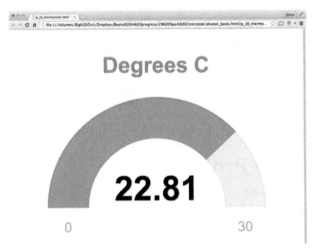

Figure 5-22. *The temperature in centigrade*

Summary

In this chapter, you have used analog inputs attached to sensors and digital outputs controlling LEDs and buzzers to make some simple IoT projects. In the next chapter, you will learn more about the Internet of Things and how to use the Photon and Core with the popular If This Then That (IFTTT) web service.

6/If This Then That

The If This Then That (IFTTT) web service allows you to link various Internet services in a simple way. IFTTT (*http://ifttt.com*) allows you to define *recipes* such as "If someone forks one of my GitHub projects, send me an email, telling me who." IFTTT also allows linking of more conventional web services like Gmail, Twitter, Facebook, and so on. IFTTT can also integrate with IoT services, including Particle. This means that you can make IFTTT recipes that use a Photon or Core. For example, "When the temperature rises over 70 degrees F, send me an email" or "Whenever anyone mentions me in a tweet, have my device ring a bell" or even "Flash the subject line of any emails I receive to me as Morse code."

If This Then That

To get started with IFTTT, you will need to visit *http://ifttt.com* and register. Once you are registered, you can get started writing your own recipes, or for that matter, use recipes that other people have contributed.

IFTTT integrates lots of other web services such as Facebook, Twitter, and Gmail; and to be able to create and use recipes that use these web services, you will have to grant IFTTT access to these accounts by logging into these separate accounts as you set up a recipe that uses them.

At this point, you might like to explore IFTTT, read their instructions on getting started, and even try out a simple recipe or two before you start using it with the Photon/Core.

Project 11. Temperature Email Alerts

This project uses the same code on the Photon/Core as Project 10. In fact, it has exactly the same hardware as Project 10 too.

So, if you have not already done so, build Project 10 and try it out before moving on to this project.

In this project, the web page that we created to show the temperature in Project 10 will not be used (although you can still use it for testing). Instead, you are going to create an IFTTT recipe that monitors the temperature Variable and sends you an email when a temperature of 70 degrees F is exceeded.

Make Project 10 and power up the Photon/Core so that IFTTT can find it, and then start by creating a new IFTTT recipe by clicking the Create a Recipe button (Figure 6-1).

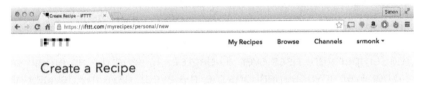

Figure 6-1. *Creating a new recipe in IFTTT*

The big clue here is the hyperlinked word *this* that is inviting you to click it so that you can define a trigger. This will take you to a screenful of icons for all the various web services that IFTTT knows about, which can act as *trigger channels*. Scroll down the list until you find the icon for Spark (Particle), shown in Figure 6-2. If this is the first time you have used Particle with IFTTT, you will be asked to log into Particle. Remember, you are already logged into IFTTT; the login details you need to enter here are the username and password for your Particle.io account, so that IFTTT can have access to it.

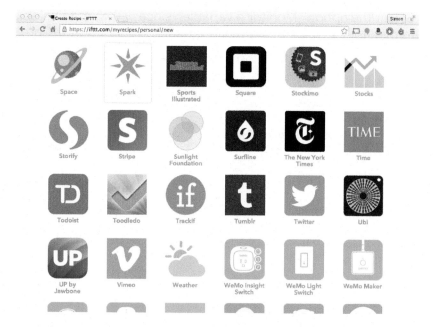

Figure 6-2. *Selecting a trigger channel in IFTTT*

Click the Spark (Particle) icon and you will be presented with a list of triggers to choose from (Figure 6-3).

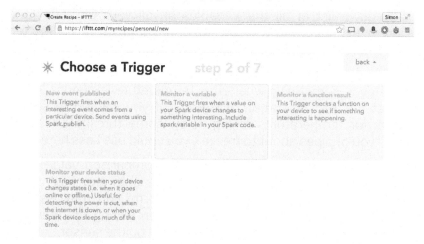

Figure 6-3. *Selecting a trigger in IFTTT*

Select the option "Monitor a variable," since you want to be able to monitor the value of the `tempF` Variable on your Photon/Core.

This will open the simple form shown in Figure 6-4.

Figure 6-4. *Configuring the trigger*

When you click the If (Variable Name) field, you will be shown a drop-down list of all the Variables on all your Photons (and Cores). In Figure 6-4, you can see that "tempf on B" is selected (my Photon is rather unimaginatively called B).

In the next field, you need to select the Test Operation. We want to trigger something when the temperature exceeds 70, so select the Test Operation of Greater. If you wanted to set up alerts for temperatures falling below some threshold (say, to warn you of pipes about to freeze), you could use Less here.

In the final field of Figure 6-4, set the Comparison Value to 70 and then click Create Trigger to move onto the next stage (Figure 6-5). If you live somewhere warm, you may want to increase 70 to perhaps 80, so that, when testing the recipe, it isn't triggered until you deliberately warm up the sensor.

tempf on "B" is Greater 70

Figure 6-5. *Defining an action in IFTTT*

Now you are being invited to complete the That part of the recipe. Notice that underneath the If part of the recipe it says "tempf on 'B' is Greater 70." Now click the That hyperlink and you will be given a selection of action channels to select from (Figure 6-6).

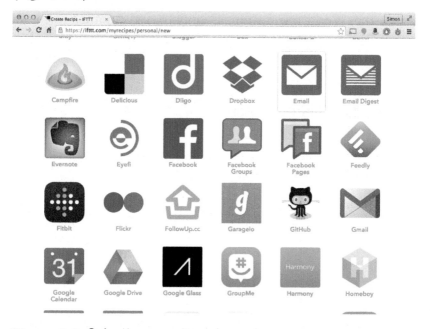

Figure 6-6. *Selecting an action channel*

The action we want is to have an email sent to us. The email will come from IFTTT's email server. IFTTT already knows your email address from when you registered, so select the Email icon from the list of action channels.

Next, you need to select the action you want to take with that channel, and there is only one choice ("Send me an email"), shown in Figure 6-7.

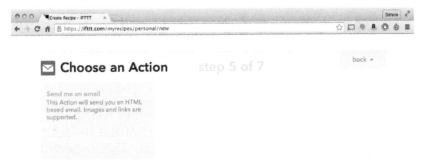

Figure 6-7. *Selecting the send email action*

Click the "Send me an email" option, and another form will appear for you to complete (Figure 6-8).

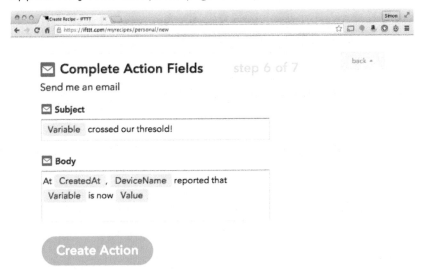

Figure 6-8. *Options for sending email*

This form will determine what will go into the email that gets automatically sent. It is a mixture of regular text and *ingredients* (Variable, CreatedAt, DeviceName, and Value). These ingredients can be used in both the subject and body of the email. Their meanings are as follows:

Variable
The name of the Variable on the Photon/Core, in this case, always `tempf`

CreatedAt
The date and time when the email was triggered

DeviceName
The name you gave to your Photon or Core

Value
The actual value of the Variable at the time that the email was triggered

Change the fields so that they look like Figure 6-9. Note that when you go to edit a field, the variables in that field will be converted to text enclosed in to mark them out as variables.

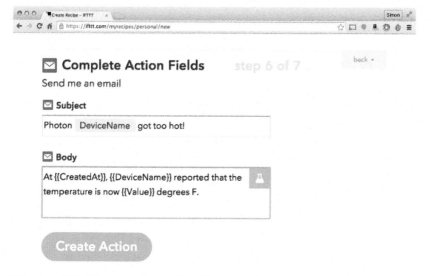

Figure 6-9. *The send-email option configured*

Click the Create Action button and you will come to the final screen, which asks you to "Create and Activate" the recipe (Figure 6-10). Once you do this, the recipe will go live! IFTTT guarantees to check only every 15 minutes, so you may not receive an email immediately and you are not going to be flooded with emails.

Create and activate   step 7 of 7   back ▲

tempf on "B" is Greater 70   Send me an email at
                                       srmonk@gmail.com

Recipe Title

If tempf on "B" is Greater 70, then send me an
email at srmonk@gmail.com

use '#' to add tags

Create Recipe

Figure 6-10. *Creating and activating the new recipe*

Also, you may need to warm up your sensor to over 70 degrees F by pinching it between your fingers, or leaning it against a warm mug of coffee (which you can drink while you wait for the email).

Eventually, you should get an email something like the one in Figure 6-11.

Of course there are all sorts of other actions that you could trigger as a notification, so you may like to go back and modify the Action part of the recipe to, say, send a tweet or Facebook status update.

In the next project, you will learn how to use a Photon or Core with IFTTT to act as an action rather than a trigger.

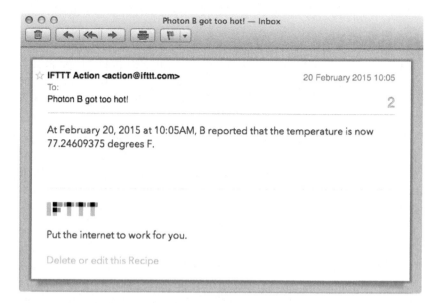

Figure 6-11. *An email notification from IFTTT*

Project 12. Ring a Bell for Tweets

This is another project that includes hardware and software that you already created in an earlier project. In this case, the project in question is Project 7, where you used a relay shield and connected a bell to one of the relays so you could turn the bell on and off over the Internet.

In this case, you will use IFTTT to ring the bell every time someone mentions your name in a tweet.

The bell used in Project 7 was kind of loud. Rather than have it sound for any longer than necessary, in this project, it will be turned on for a fraction of a second—just long enough for it to alert you, without driving you insane.

You could, of course, wire up a lightbulb or some other device to the relay if you want something a bit quieter.

Software

The program for the Photon/Core is similar to that of Project 7, so please refer back to that project for a full description of the code. You can find the code in the files *p_12_tweet_bell* in the *PHOTON_BOOK* library examples.

The program is still general purpose, in that it allows you to control any of the relays. However, the Function name has changed to `relaycontrol`, and an extra option of specifying an "on" time has been added. This will allow us to make the duration of the bell sounding just short enough to alert us to the tweet.

The syntax of the command is backward compatible with the old relay command, so that you can still use the command 31 to turn relay 3 on, and 30 to turn relay 3 off again. However, now there is a different, separate command P (for pulse) in place of 1 and O for on and off that then takes another parameter of the duration in milliseconds for the relay to be on. For example, 3P-1000 will turn relay 3 on for one second.

The code for the `relaySwitcher` function that implements this is listed here:

```
int relaySwitcher(String command) {
    // "11", "10", "1P-1000" - P for pulse duration in millis
    int relayNumber = command.charAt(0) - '0';
    char action = command.charAt(1);
    if (action == '1') {
        digitalWrite(relayPins[relayNumber-1], HIGH);
    }
    else if (action == '0') {
        digitalWrite(relayPins[relayNumber-1], LOW);
    }
    else if (action == 'P' || action == 'p') {
        int duration = command.substring(3).toInt();
        digitalWrite(relayPins[relayNumber-1], HIGH);
        delay(duration);
        digitalWrite(relayPins[relayNumber-1], LOW);
    }
    return 1;
}
```

The new part of the function (the final `else if` clause) first checks to see if the action is P or p and then uses a substring from position 3 to the end of the command and converts this from a string to an integer by using `toInt`.

Turning the relay on for the right duration is then just a matter of setting the appropriate relay pin high, delaying for the required number of milliseconds, and then turning it off again.

Flash the program onto your Photon/Core so that IFTTT will be aware of the new action.

IFTTT

Log in to your IFTTT account and then, when choosing a trigger channel, select Twitter. If it's the first time that you have used Twitter from IFTTT, it will ask you to authenticate with Twitter so IFTTT can monitor your Twitter account. There are quite a few triggers that you can select from within Twitter. The one I decided on was "New Mention of You."

Select "New Mention of You," and the next screen will tell you that there are no fields to complete, so just click Create Trigger. You can now move on to the That section of the recipe.

Click the Spark (Particle) icon in the Action channel source and you will be given the choices shown in Figure 6-12.

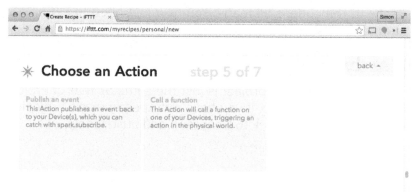

Figure 6-12. *Selecting an action*

Select the option "Call a function" and you will be asked to Complete Action Fields. From the "Then call" drop-down, select "relay control" on whatever Photon or Core you just flashed, and then delete the contents of the "with input" field with the text "1P-100" (see Figure 6-13).

Figure 6-13. *Configuring an action*

Click Create Action and then click Create Recipe.

That's it; the code is all ready. You now just need to set up the hardware.

Hardware

The hardware for this project is just the same as that of Project 6 (see Figure 6-14). So, if you have not already made Project 6, you might like to go back to Chapter 5 and build it now.

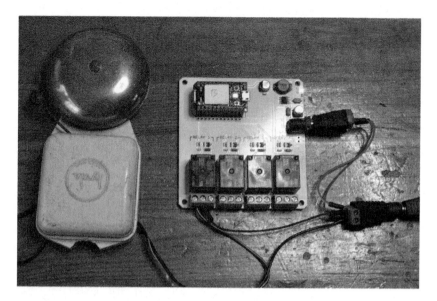

Figure 6-14. *The hardware from Project 6*

Using the Project

To test the project, you need to mention yourself in a tweet, or get someone to mention you. When IFTTT notices this, the bell will ring. You can adjust the duration of the bell ring by editing the parameter to the action in the IFTTT recipe.

Note that IFTTT guarantees to check triggers only every 15 minutes, so you will probably have to wait a minute or two before you know if the project worked.

An alternative to alerting you of mentions would be to alert you of new followers or when you tweet using a specific hashtag. Or, you could chose another type of trigger entirely, such as sending an email (use the Email trigger source) or simply at a certain time of day using the Date & Time trigger source.

Project 13. Flash Email as Morse Code

The final IFTTT example project takes the Morse code flasher of Project 8 and gives it an IFTTT twist. The twist is that when you send an email to IFTTT, the subject of that email will be flashed as Morse code.

Software

The code running on the Photon/Core is exactly the same as for Project 8. Just flash your Photon with the program *p_08_Morse_Function* from the *PHOTON_BOOK* library.

Hardware

The hardware for the project is also exactly the same as for Project 8. The software will flash the built-in LED on D7, so if you do not want to use an external LED or buzzer, then you can just use the Photon/Core on its own without any components attached to it at all.

If you have decided that you want a buzzer and external LED, follow the instructions for Project 8 in Chapter 6.

IFTTT

To create the IFTTT recipe for this project, create a new recipe and then select Email as the trigger source. In IFTTT, the Email trigger source means emails to and from the IFTTT email server. In this case, select the trigger "Send IFTTT any email." There are no fields to enter, so at the next screen click Create Trigger.

The action is more interesting. Select Spark (Particle) as the action channel and then the "Call a Function" action.

In the "Then call (Function Name)" field, select the Function morse on whatever Photon/Core you have just flashed the program onto.

Now you need a way of passing the subject of the email to the morse Function. To do this kind of thing, IFTTT provides the concept of an ingredient for the recipe. Delete the contents of the "with input (Function Input)" field and click the beaker icon on the right. The pop-up list in Figure 6-15 will then appear, and you can click the Subject ingredient.

✳ Complete Action Fields

Call a function

✳ Then call (Function Name)

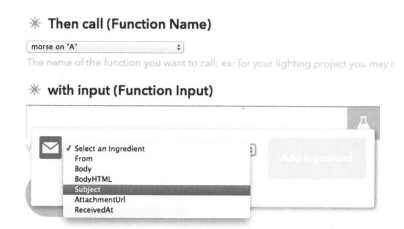

Figure 6-15. *Selecting ingredients for your recipe*

This will insert the text into the field. Click Create Action and then Create Recipe.

Using the Project

To test the project, send an email to *trigger@recipe.ifttt.com* with the subject line of the text that you want to be flashed and beeped as Morse code. This email must be sent from the email account that you used to register with IFTTT.

Summary

In the next chapter, you will learn how to use a Photon/Core to create a small roving robot that is controlled from a web page and that also displays its battery voltage and the distance to any obstacle in front of it.

7/Robotics

The Photon's small size and wireless connectivity makes it a great choice for robotics projects. In this chapter, you will learn how to use a Photon or Core with a PhoBot shield to control a small roving robot. The robot is also fitted with an ultrasonic range finder module to help prevent it from bumping into things.

Project 14. Web-Controlled Robot

This project makes use of a robot controller board called the PhoBot by MonkMakes. You may guess by the company name that I had something to do with the design of this product, which you can buy from several sources (see Appendix A).

The board is combined with a chassis kit that provides a base and gearmotors to drive the robot, as well as wheels and a battery pack.

Figure 7-1 shows the finished rover, and Figure 7-2 shows the web page to control it and provide feedback about the state of the rover's battery and the distance to any obstacles in front of it.

Figure 7-1. *A Photon roving robot*

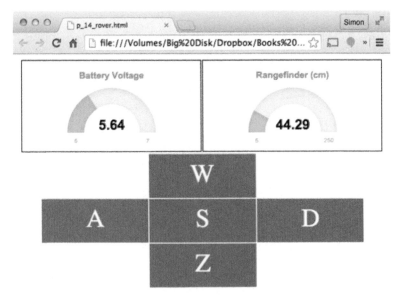

Figure 7-2. *A web page to control the rover*

Parts

To build this project, you need the parts listed in Table 7-1 in addition to your Photon/Core.

Table 7-1. *Project 14 parts bin*

Part	Description	Appendix A code
PhoBot shield	PhoBot robot controller shield	M2
Rangefinder	HC-SR04 ultrasonic rangefinder	M3
Chassis kit	Rover robot chassis kit (6V)	H6
Hookup wire	Wire to connect the motors to the PhoBot shield	H3

There are many low-cost robot chassis on the market, and most will work with the PhoBot. Look for a kit with 6V motors, and a 4 x AA battery holder, although lower-voltage motors will also work.

You will need the hookup wire only if the gearmotors supplied with the chassis kit do not have wires attached. In that case, you will need to solder wires about 6 inches long to each motor terminal.

Software (Photon)

It's a good idea to program your Photon/Core with the app for this project before attaching it to motors, where some inadvertent use of output pins on a previous project could have your rover driving off the end of your table.

The program for this project is the example program provided in the PhoBot library that makes the PhoBot easy to use. Find the *PHOBOT* library by typing PHOBOT into the search field below Community Libraries in the Web IDE.

When you do this, you will see the library files. Select the file *WEBROVER.CPP* and then click the USE THIS EXAMPLE button. This will create a copy of the example program that we can use to program the Photon/Core, but can also be modified if we want to change things a little.

Before the *WEBROVER* example can be flashed onto a Photon or Core, you will need to find a couple of libraries that are used by the *PHOBOT* library and also the *HC_SRO4* library needed for the rangefinder. If you are using a Spark Core rather than a Photon, you will also need to import the *SPARKINTERVALTIMER* and *SOFTPWM* libraries. To import each of these libraries for use with the example *WEBROVER*, you must find each library (*HC_SRO4*, *SPARKINTERVALTIMER*, and *SOFTPWM*) in turn in the Community Libraries area and then click the INCLUDE IN APP button. When the Web IDE prompts with "Which app," select *WEBROVER* from the list and click the ADD TO THIS APP button. Repeat this for all three libraries. You will notice that each time that you do this a new #include statement will appear at the top of the program.

The libraries take care of most of the code for the rover for us. Here is the listing for the *WEBROVER*:

```
#include "HC_SRO4/HC_SRO4.h"
#include "SparkIntervalTimer/SparkIntervalTimer.h"
#include "SoftPWM/SoftPWM.h"
#include "PhoBot/PhoBot.h"

double volts = 0.0;
double distance = 0.0;

PhoBot p = PhoBot();
HC_SRO4 rangefinder = HC_SRO4(p.trigPin, p.echoPin);

void setup() {
    Spark.function("control", control);
    Spark.variable("volts", &volts, DOUBLE);
    Spark.variable("distance", &distance, DOUBLE);
}

void loop() {
    volts = p.batteryVolts();
    distance = rangefinder.getDistanceCM();
}

int control(String command) {
    return p.control(command);
}
```

Note that the extra comments added next to each line (including some that appear automatically after using a library) have been removed for clarity.

This program uses two Variables: `volts`, which contains the battery voltage, and `distance`, which contains the last rangefinder measurement in centimeters.

After the two `double` variables, the PhoBot and rangefinder are initialized and assigned to the variables `p` and `rangefinder`, respectively. When the rangefinder is initialized, the pins for it to use are taken from the PhoBot's library as `p.trigPin` and `p.echo Pin` (see the following sidebar).

Ultrasonic Rangefinders

Ultrasonic rangefinders work by timing how long it takes for a pulse of ultrasound to bounce off an obstacle and return to the sensor (Figure 7-3). Using the speed of sound, the distance is calculated.

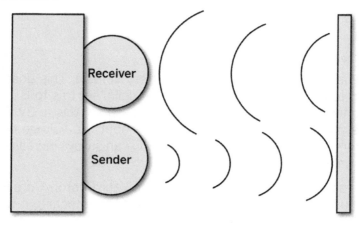

Figure 7-3. *How an ultrasonic rangefinder works*

The HC-SR04 module used for this has power pins and two additional pins, trigger and echo. When the trigger pin is taken high for a fraction of a second, the module emits a pulse of ultrasound. When that pulse returns, the echo pin will indicate its return.

The **setup** function defines a Function called **control**. The **con trol** function uses high-level commands like **F-100**. The first letter is the direction that the bot is to travel in (F—forward, B—backward, L—turn left, R—turn right). The parameter after the hyphen is the speed, where 100 is full speed, 50 would be half speed, etc. The command S without any speed parameter stops the bot.

The two Variables are also defined in setup: **volts** and **distance**.

The **loop** function updates both variables with the current battery voltage and distance to an obstacle.

The function **control** is the handler for the **control** Function and passes on whatever command it is given to the function of the same name within the PhoBot library.

One tweak to mention here is that if your chassis has motors of a lower voltage than that of the battery pack (say, 3V), you can change the code slightly to compensate for this. Replace the following line of code:

```
PhoBot p = PhoBot();
```

with

```
PhoBot p = PhoBot(6.0, 3.0);
```

The first parameter (6.0) is the battery voltage. This does not need to be exact, if you have four AA cells, set this to 6.0. The second parameter is the motor voltage, so set this to 3V if you have 3V motors. If you are not sure about the voltage of the motors, start by setting them to 3V. You can always increase this if it makes the motors run too slowly.

You can flash the *WEBROVER* app onto your Photon/Core while it is just connected to USB before you start attaching it to the hardware and its battery pack.

Software (Web Page)

The web page for this project uses a combination of techniques that you have already seen in earlier projects and is too long to list here, so you might want to open the file *p_14_Rover.html* in a text editor.

This project uses three URLs, for the two Variables and the control Function:

```
var accessToken = "cb8b348000e9d0ea9e354990bbd39ccbfb57b30e";
var deviceID = "55ff6b0650755555322151487"
var volts_url = "https://api.spark.io/v1/devices/"
                + deviceID + "/volts";
var distance_url = "https://api.spark.io/v1/devices/"
                + deviceID + "/distance";
var control_url = "https://api.spark.io/v1/devices/"
                + deviceID + "/control";
```

Remember that you will need to change the accessToken and deviceID for your account details.

Since there are now two Variables to be displayed in gauges, there are two corresponding callback functions to update the gauges, callbackVolts and callbackDistance. Both are much the same as earlier examples. The callback for the battery voltage finishes by scheduling another reading after 10 seconds, rather than the 1-second update of the rangefinder reading, as the battery voltage shouldn't change that quickly.

The web page (see Figure 7-2) has buttons to control direction, but also maps key presses on the keyboard to the various direction commands. The following line is responsible for intercepting key presses while you are on this web page and calls the function keyPress every time a key is pressed:

```
$(document).keypress(keyPress);
```

The keyPress function is listed here:

```
function keyPress(event){
    code = event.keyCode;
    if (code == 119) {
        sendControl('F-100');
    }
    else if (code == 97) {
        sendControl('L-50');
    }
    else if (code == 115) {
        sendControl('S');
    }
    else if (code == 100) {
        sendControl('R-50');
    }
```

```
    else if (code == 122) {
        sendControl('B-75');
    }
}
```

Each key has a character code, based on the ASCII code for that letter. If you search the Internet you will find information about ASCII codes for letters. In this case, *w* is 119, *a* is 97, *s* is 115, *d* is 100, and *z* is 122. When a particular key is pressed, the relevant command is sent to the Photon/Core.

One feature of the code for the volts gauge is that the gauge should go red as the battery becomes exhausted. This is achieved using a color list associated with the gauge. The three colors are red, orange, and green as # hex strings.

```
var volts_gauge = new JustGage({
    id: "voltsGauge",
    value: 0,
    min: 4,
    max: 7,
    levelColors: ["#FF0000", "#FFFF00", "#00FF00"],
    title: "Battery Voltage"
});
```

Hardware

Before you start attaching any electronics, you will need to assemble the robot chassis. Most kits will comprise the following items:

- A laser-cut acrylic plastic base
- Two gearmotors (ideally 6V)
- Two wheels to fit the gearmotors
- A battery box to fit from four to six AA battery cells
- A castor or universal wheel to be fitted at one end of the chassis

Before you assemble the chassis, you may need to solder leads to the gearmotors that are long enough to reach around to the screw terminals of the PhoBot board.

Once the chassis is assembled, work out where you want to place everything on the chassis. The heaviest part of the rover will be the battery holder (once it has batteries in it), so make sure that this is positioned inside the triangle formed by the castor and the two drive wheels. Otherwise, the rover might tip over.

Remember that the rangefinder will need a clear view in front of it to be able to detect obstacles. If you are lucky, screw holes will be in the right place, and you can screw the PhoBot shield and battery box to the chassis base. If not, then self-adhesive Velcro^{fi} pads are a great way of attaching things.

Figure 7-4 shows the top view of the wired-up rover. The rangefinder has been temporarily removed for a clearer view of the screw terminals.

Figure 7-4. *Wiring the PhoBot*

The battery box should have one red (+) and one black (−) lead. The red lead should go to the screw terminal labelled 6V, and the black lead from the battery box is connected to the screw terminal marked GND.

The PhoBot shield can control up to four motors, but only the pairs of screw terminals marked M3 and M4 are capable of bidirectional operation (making the motor run both clockwise and counterclockwise), so we will use these rather than M1 and M2.

Thread the leads from the motors through a convenient hole in the chassis base. Connect the lefthand motor leads to the pair of screw terminals marked M4 and the righthand motor leads to the pair marked M3.

You will not really know if these leads for a particular motor are the right way around until you try to drive them. Reversing the leads on a DC motor causes it to rotate in the opposite direction.

Now you can plug the Photon/Core and rangefinder onto the PhoBot shield. Make sure that they are both the right way around. You can also fit the batteries into the battery holder now.

If the battery holder does not have a switch, pull one end of one of the batteries out of the battery box as a crude switch.

Using the Project

Chances are you are going to need to swap over a few of the motor wires, so it's best to get all of that sorted out while the wheels are still off the motors. That way, you can sit the rover in front of you without worrying about it wandering off.

Open *p_14_rover.html* in your web browser and press the F button (or key) and watch what the motor shafts do for moving forward. One of the following situations will apply:

- They are both moving correctly. Hurray, you are finished!
- One of the motors is rotating in the wrong direction. If so, swap over the wires for that motor only, or both motors if both are wrong.

Now press the A button or the A key on your keyboard and watch what the motors do. If you performed the preceding step correctly, the motors will be moving in opposite directions. The question is whether this is turning you to the left (counterclockwise) or right (clockwise). If the answer is counterclockwise, all is well and you are done. Otherwise, you will need to swap the motor that is connected to M3 to be connected to M4, and vice versa. When doing this, keep the leads for each motor in the same order (for example, red on the left, black on the right).

The rover is fairly power-hungry, so when it's not in use, pull one end of one of the batteries out to turn it off.

If you are using rechargeable batteries, these are lower voltage than single-use AA batteries (generally 1.2V rather than 1.5V). If the battery voltage falls too low, the Photon/Core will crash, and you will lose control of the motors. Using a five- or six-cell battery holder will give you more drive time between charges if you are using rechargeable batteries. For single-use batteries, four AA batteries should give you a couple of hours of driving time.

A nice refinement to the code might be to monitor the battery voltage and automatically stop the rover when the batteries get low.

Summary

A roving robot like this can form the basis for all sorts of interesting projects. You could, for instance, attach a wireless webcam to it and turn it into a surveillance robot.

In Chapter 8, you will learn about Particle's powerful publish-and-subscribe model and how to use it when making two or more Photons or Cores communicate with each other.

8/Machine-to-Machine Communication

Most of the projects in this book rely on the Photon or Core either being completely independent (once programmed) or being controlled through a web browser or other user interface over the Internet. In this chapter, you will learn how to make one Photon/Core speak to another, or even many other Photons or Cores.

Publish and Subscribe

In the previous chapter, when you used IFTTT to monitor the temperature on a Photon/Core, IFTTT checked the temperature Variable on the Photon only every 15 minutes. The reason they limit it in this way is that otherwise there could be potentially huge volumes of web and email traffic if everyone's triggers were being monitored continuously. This type of checking on values is called *polling* and is not very efficient.

The alternative to polling in this situation would be to move the temperature testing to the Photon/Core, so that only when the level is exceeded would the device initiate communication by *publishing* a temperature-exceeded event to any services such as IFTTT or other Photons that are *subscribing* to that event.

In other words, a device or web service can subscribe to an event. That is, they can register an interest in that event, so that when that event is published, they are notified. This is a lot more efficient than continually polling and means that you don't have to (in the case of IFTTT) wait for 15 minutes before you receive your notification.

Temperature Monitor Example

If you have made Projects 10 and 12, and have two Photons, then you can try out this project using the exact same hardware as those two projects. You will just need to flash the new software onto the Photons.

Figure 8-1 shows how the publish-and-subscribe model can be used to create a network of Photons, all communicating with each other.

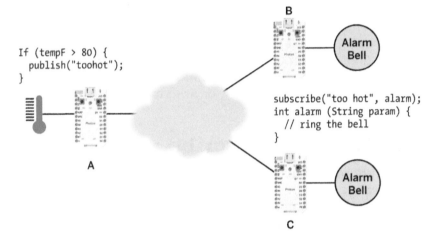

Figure 8-1. *Publish and subscribe*

In this case, there is one Photon acting as a thermometer (A), and there are two Photons with alarm bells attached (B and C). The intention is that these alarm bells will ring whenever Photon A publishes a **toohot** event. Previously (in their **setup** functions), Photons B and C will have registered an interest in the **toohot** notification by subscribing to **toohot**. Once they do this, the Particle cloud service knows that every time Photon A says it's too hot, Photons B and C need to be informed.

The code on Photon A will look something like the following code. This code is a little flawed. You will fix this later, so, for now, don't flash this onto your Project 10 hardware. Wait until it has been improved to solve excessive publishing.

```
#include "spark-dallas-temperature/spark-dallas-temperature.h"
#include "OneWire/OneWire.h"
```

```
int tempSensorPin = D2;
OneWire oneWire(tempSensorPin);
DallasTemperature sensors(&oneWire);

void setup() {
    sensors.begin();
}

void loop() {
  sensors.requestTemperatures();
  float tempC = sensors.getTempCByIndex(0);
  float tempF = tempC * 9.0 / 5.0 + 32.0;
  if (tempF > 80.0) {
      Spark.publish("toohot");
  }
}
```

Most of this code is concerned with measuring the temperature,
so refer back to Project 10 for more information on that topic.
The line with the **publish** call shows just how easy it is to publish
an event.

The code for Photon B (the relay and bell hardware) is shown
here:

```
int relayPin = D0;
void setup() {
    pinMode(relayPin, OUTPUT);
    Spark.subscribe("toohot", soundAlarm);
}

void loop() {
}

void soundAlarm(const char *event, const char *data) {
    digitalWrite(relayPin, HIGH);
    delay(200);
    digitalWrite(relayPin, LOW);
}
```

The **setup** function contains the code to subscribe to the **toohot**
event and specifies that the function **soundAlarm** should be
called if such an event happens. You will notice that **soundAlarm**
has some strange parameters; you'll learn how these work later.

Although this pair of programs will work, there is (as I mentioned earlier) a slight flaw in the first program. The temperature is checked in the **loop** function, and just so long as the temperature is over 80, the event will be published every time around the loop. So Photon A will be publishing the **toohot** event over and over again, as fast as it can, while the temperature is over 80. This is not very socially responsible programming. It would be better if we used two events. One event would be published when the temperature exceeded 80, and a second event (**tempnormal**) would be published when the temperature had fallen back below, say, 78. The reason for a second threshold temperature of 78 rather than 80 is that if both threshold temperatures were 80, then at the point where the actual temperature was just around 80, the readings could alternate between 79 and 80, causing a flurry of both events.

The modified code for the thermometer end (Photon A) is listed here:

```
#include "spark-dallas-temperature/spark-dallas-temperature.h"
#include "OneWire/OneWire.h"

int tempSensorPin = D2;
OneWire oneWire(tempSensorPin);
DallasTemperature sensors(&oneWire);

boolean toohot = false;

void setup() {
    sensors.begin();
}

void loop() {
  sensors.requestTemperatures();
  float tempC = sensors.getTempCByIndex(0);
  float tempF = tempC * 9.0 / 5.0 + 32.0;
  if (tempF > 80.0 && toohot == false) {
      Spark.publish("toohot");
      toohot = true;
  }
  if (tempF < 78.0 && toohot == true) {
      Spark.publish("tempnormal");
      toohot = false;
```

```
        }
    }
```

A new **boolean** variable **toohot** has been added to the program, so now, after reading the temperature, the **toohot** event is published only if the variable **toohot** is **false**, and the measured temperature has risen above 80. The variable **toohot** is then set to **true**, to prevent any further publishing of **toohot** until the temperature has fallen back below 78.

The second **if** statement takes care of this. If the measured temperature is below 78 and the **toohot** variable is currently **true**, then the **tempnormal** event is published and **toohot** set to **false**.

The code for the relay end of the project (Photon B) also needs to be modified to handle the new event (**tempnormal**):

```
int relayPin = D0;
void setup() {
    pinMode(relayPin, OUTPUT);
    Spark.subscribe("toohot", soundAlarm);
    Spark.subscribe("tempnormal", cancelAlarm);
}

void loop() {
}

void soundAlarm(const char *event, const char *data) {
    digitalWrite(relayPin, HIGH);
}

void cancelAlarm(const char *event, const char *data) {
    digitalWrite(relayPin, LOW);
}
```

Now, **setup** subscribes to both **toohot** and **tempnormal** events. The **toohot** event will turn the alarm bell on, and **tempnormal** will turn it off again—just what we want.

If you want to try out these programs, the thermometer program is in the file *ch_08_Temp_monitor_Pub* and the relay program in *ch_08_Temp_monitor_Sub*. These will run on the hardware for Projects 10 and 12, respectively.

IFTTT and Publish/Subscribe

IFTTT is able to use the Particle publish-and-subscribe system. For example, you could arrange for an email to be sent when the temperature is exceeded not by monitoring a Variable as we did in Project 10, but by having an IFTTT recipe that subscribes to the **toohot** event, which would be much more efficient.

Create a new recipe in IFTTT and then select Particle as the trigger source. Instead of selecting "Monitor a Variable," select "New event published" as the trigger. Figure 8-2 shows the Complete Trigger Fields page.

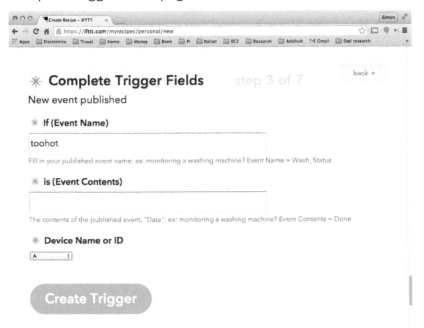

Figure 8-2. *Setting an event trigger*

Specify **toohot** as the event name and leave the Event Contents field blank. Finally, select the Photon/Core that is going to be sending the events.

Select an action channel of Email and then the action of "Send me an Email," as you did in Project 11.

When the recipe is finished and active, you will receive just one email each time the event is triggered. Pinch your temperature sensor between your fingers to warm it up and test this example.

Advanced Publish and Subscribe

The preceding example uses publish and subscribe at its simplest, and for many applications that will be all that you need. However, the Particle framework for publish and subscribe is capable of more advanced features.

Publish

You can read the full documentation for the `publish` command at *http://docs.spark.io/firmware/#spark-publish*.

In the preceding example, the only parameter that we supplied to `publish` was the name of the event. In fact, you can supply the following additional parameters:

data
> This string could contain a a value to accompany the event. For example, the actual temperature could be sent, but this would need to be converted to a string first.

time to live
> This value specifies the number of seconds that the event should be allowed to live before it is automatically removed. This prevents too many events accumulating in the system. At the time of writing, this parameter is ignored by the Particle cloud, which discards events after the default of one minute.

public/private
> The default is public, meaning that anyone can subscribe to these events. This allows for the possibility of interesting collaborative projects.

Subscribe

The `subscribe` command also has an extra parameter that follows the name of the handler function. This specifies the scope of the events that are to be subscribed to. You can use this third parameter to limit subscriptions to events coming from a particular Photon or Core by supplying the device's ID, like this:

```
Spark.subscribe("toohot", soundAlarm,
                "55ff700649555344339432587");
```

You can also limit the subscription to just your own devices by using the following command:

```
Spark.subscribe("toohot", soundAlarm, MY_DEVICES);
```

Subscription name matching is even more subtle than this. It will match against however much of the name you supply, so the following line will still receive the **toohot** events:

```
Spark.subscribe("too", soundAlarm);
```

This opens up the possibility of organizing events into related families based on the event name. Note that the maximum size of an event name is 64 characters.

Project 15. Magic Rope

This project was inspired by a video made by Leena Ventä-Olkkonen, Tobi Stockinger, Claudia Zuniga, and Graham Dean that showed how a public installation could be made that would allow large maps of the world to be positioned in various public spaces in a city. These maps would have short lengths of rope sticking out of holes on the map corresponding to other cities around the world. The idea is that the public at any one of these cities (let's say London) could walk up to a rope at their map, pull on it, and the paired rope in the other city (say, New York) would be pulled into the map, attracting the attention of people near the installation. A gentle exchange of rope pulling could then occur across the world. You can view the original video at *http://bit.ly/1aex7Jk*.

The original project was developed only as a concept and not actually implemented as a real installation. In this project, you will make a pair of "entangled" ropes that could be positioned in different cities. This could be used as a nice way of staying in touch with distant relatives.

Figures 8-3 and 8-4 show the pair of Photons with their tempting lengths of string to be pulled. One is in a wooden box, adding to the mystery.

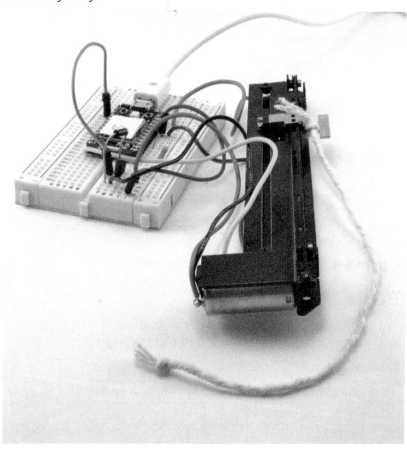

Figure 8-3. *The Magic Rope project naked*

Figure 8-4. *The Magic Rope project in a box*

Parts

To build this project, you need two sets of each of the parts listed in Table 8-1 in addition to two Photons/Cores.

Table 8-1. *Project 15 parts bin*

Part	Description	Appendix A code
R1	Slide pot—motorized	M4
R2	220Ω resistor	C1
Q1	2N3904 transistor	C10
D1	1N4001 diode	C11
	Male-to-male jumper wire	H4
	Half-sized breadboard	H5
	Assorted hookup wire	H3

The sliding pots (potentiometers) used in this project are variable resistors intended for use in automated music-mixing desks. You can adjust the resistance by sliding the "slider" up and down the length of the pot, but there is also a little motor that will move the slider by using a toothed belt drive.

These motorized pots do not have leads attached to the pins, so this is one project where you will need to use a soldering iron and attach some leads to the pins.

Software

Both ends of this project have exactly the same software running on them, and you can find it in the file *p_15_Magic_Rope* in the *PHOTON_BOOK* library:

```
int motorPin = D4;
int potPin = A0;

String thisID = Spark.deviceID();
boolean myTurn = true;
int maxPosn = 4000;
int minPosn = 3000;

void setup() {
    Spark.subscribe("pulled", remoteRopePulled);
    pinMode(motorPin, OUTPUT);
    moveSliderTo(maxPosn);
}

void loop() {
    int newLocalPosition = analogRead(potPin);
    if (newLocalPosition <  minPosn && myTurn) {
        Spark.publish("pulled", thisID);
        myTurn = false;
    }
}

void remoteRopePulled(const char *event, const char *data)
{
    String dataS = String(data);
    // ignore messages from yourself
    if (dataS.indexOf(thisID) == -1)
    {
        moveSliderTo(maxPosn);
```

```
                myTurn = true;
        }
    }

    void moveSliderTo(int newPosition) {
        while (analogRead(potPin) < newPosition) {
            digitalWrite(motorPin, HIGH);
        };
        digitalWrite(motorPin, LOW);
    }
```

The original version of this file has some extra commands commented out that can be used to debug the project if the events don't seem to be getting through. See the comments in the original program if you need to use them.

The program starts by defining the two pins to be used. A0 is for the voltage output of the potentiometer, which will be 0V if the rope is fully pulled out, and 3.3V if the rope is fully pulled in.

The boolean variable myTurn is used to keep track of whose turn it is to pull on the rope. If myTurn is set to true, then it is this device's turn to have its rope pulled.

Both ends of this project both publish and subscribe to the same event, so the variable thisID is needed so that the device knows its own ID and can disregard its own publish events, reacting only to events coming from the other Photon/Core.

The constant maxPosn is the analog input reading at which the sliding pot is at the position where the rope is fully pulled in. This is set slightly lower than the theoretical maximum analog input value of 4095 to allow for any inaccuracy in the analog readings.

The second constant, minPosn, is equivalent to about three-quarters of the way pulled in, and this is the threshold at which a "pulled" event will be published.

The setup function makes the necessary subscription to "pulled" associating it with the function remoteRopePulled. It also calls the function moveSliderTo to position the slider at its fully pulled-in position, ready to be pulled out.

The `loop` function reads the analog input to find the `newLocalPo` `sition`. If this is less than the `minPosn` constant and it's this device's turn to move, then the "pulled" event is published with this device's ID as its parameter.

In the situation where the rope has been pulled on the other Photon/Core, the function `remoteRopePulled` will be called. This function will be supplied with the ID of the device where the rope was pulled, so that it can be compared with `thisID`, the ID of the receiving Photon, by searching for the string of characters in `thisID` within the ID passed in data.

If the event has come from a remote Photon/Core, the slider is pulled fully in, and `myTurn` is flipped over to `true`.

The function `moveSliderTo` handles all automated movement of the slider. In fact, it can only pull the slider *in*. But you can't push rope, so that's fine. The function takes the new position as a parameter and keeps power supplied to the motor until such time as the measured position is no longer less than the desired position.

Hardware

The breadboard layout for this project is shown in Figure 8-5.

The motorized pots are actually stereo devices, but we need only one channel for this project. This means that there are some pins that you do not need to connect leads to. Figure 8-6 shows the underside of the motorized pot. You can see the motor at the bottom right.

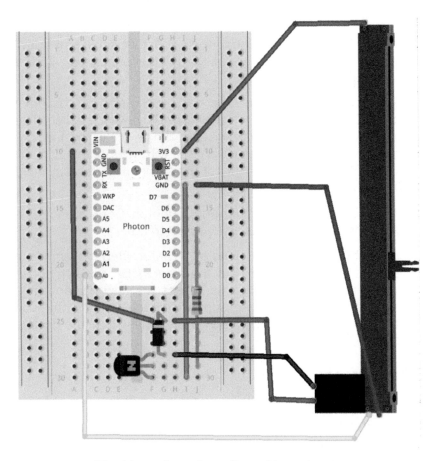

Figure 8-5. *The Magic Rope breadboard layout*

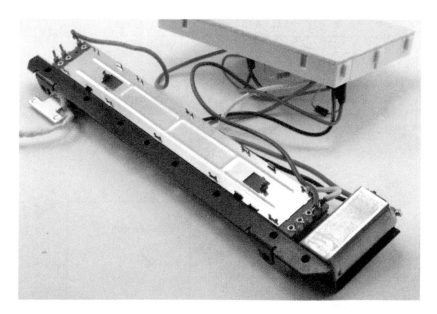

Figure 8-6. *The underside of the motorized pot*

Before assembling the breadboard, you will need to solder some wires to the motorized pot legs. The motor leads are easily identified. If you are using the same motorized pot as I am, attach a red lead to the bottommost motor lead (as shown in Figure 8-6) and a black lead to the other motor lead. All leads need to be about 6 inches long to comfortably reach the breadboard.

At the far end of the motor, attach a red lead to the rightmost lead. This is the lead that will go to 3.3V on the breadboard. Attach a yellow or orange lead to the rightmost pin at the motor end of the pot. This lead is the slider of the pot that will connect to A0 on the Photon/Core. Finally, connect a brown or blue lead next to this yellow lead. This will connect to GND on the breadboard.

Transistors

The motors of the motorized pots use upwards of 100mA. This is far too much for a Photon/Core digital output, which has an upper limit of about 20mA. To allow the digital output to turn the motor on and off, a transistor is used.

Figure 8-7 shows the schematic diagram for the project.

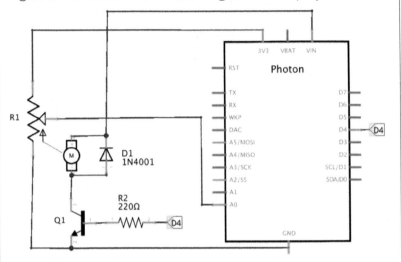

Figure 8-7. *Using a transistor to control a motor*

You can think of the transistor (Q1) as a kind of digital switch that can use a small current to control a much bigger current.

When D4 is HIGH, a small current flows through R2 and the transistor to GND, which stimulates a much higher current to flow from VIN (5V), through the motor, through the transistor, and to GND.

Finally, connect everything as shown in Figure 8-5, paying special attention to the transistor and diode, to make sure they are the right way around. The diode has a stripe at one end that should be toward the top of the breadboard, and the transistor has one curved side.

Driving motors can result in voltage spikes, and the diode protects the Photon/Core from accidental damage resulting from these spikes.

Using the Project

To use the project, power up both ends of the magic rope. After the Photon/Core has finished starting up (flashing green light), the motor should activate on both ends of the project, pulling the slide up to one end of the track.

Pull on one of the strings and then the other. When you pull on the second string, the first string should be pulled back automatically.

You could find a nice wooden box for this project, drilling a hole for the string to emerge at one end and a hole for the USB lead at the other.

Summary

Use of publish and subscribe is very powerful, and opens up all sorts of possibilities for collaborative projects where people can interact physically over the Internet.

This is the final chapter containing projects. In the next chapter, various more advanced topics will be covered.

9/Advanced Photon

In this chapter, you will learn about some of the more advanced features of the Photon and Core.

Configuring a Photon Using USB

Although the Tinker app is a very convenient way of setting up a new Photon/Core, it is not the only way of telling a new device your WiFi credentials. The USB socket on a Photon or Core is not just for supplying the device with power; it can also be used to communicate with your computer over USB.

You can pass WiFi credentials to your Photon/Core by using the USB connection and a serial communication program. If you have a Mac or Linux computer, you already have a built-in utility (called *screen*) for this kind of serial communication over USB.

Windows Users

If you are a Windows user, you will need to install Putty (*http://www.chiark.greenend.org.uk/~sgtatham/putty/*) and a USB driver for the Photon/Core (*https://s3.amazonaws.com/spark-website/Spark.zip*).

Extract the driver archive (*Spark.zip*) to your desktop. You will need to point the Found New Hardware Wizard at this folder later. Plug the Photon/Core in and, when prompted for a driver, navigate to the folder you just downloaded.

To use Putty to communicate over USB with your Photon/Core, start Putty up and then, in the row of radio buttons for "Connection type," select the option Serial (Figure 9-1).

Figure 9-1. *Setting up Putty*

The Photon/Core will probably be allocated to COM7.

If you are using Linux or a Mac, open a terminal window and issue this command:

```
$ screen screen /dev/cu.usbmodem1451 9600
```

When selecting the device (i.e., cu.usbmodem1451), the number on the end may be different for you; so press Tab after you have typed *cu.usbmodem*, and your device should be autocompleted.

Whether you are using Putty or screen, you should now have an empty screen waiting for a command. Press the I key and you should see a message like this:

```
Your core id is 54ff6f0655572524851401167
```

Make a note of this number. This is the ID for your Core. You will need it later when you come to manually claim your Core.

The other command that you can send to the Photon/Core is W. This lets you tell the Photon/Core what your wireless network is called and its password. So, press the W key and you will be prompted for the SSID, security type (probably WPA2), and then your password:

```
SSID: mymobilenetwork
Security 0=unsecured, 1=WEP, 2=WPA, 3=WPA2: 3
Password: mypassword
Thanks! Wait about 7 seconds while I save those credentials...

Awesome. Now we'll connect!

If you see a pulsing cyan light, your Spark Core
has connected to the Cloud and is ready to go!

If your LED flashes red or you encounter any other problems,
visit https://www.spark.io/support to debug.

    Spark <3 you!
```

If all is well, your Photon/Core should reboot itself and then connect itself to the Particle cloud indicated by a rhythmic cyan breathing of the RGB LED.

Although your Photon/Core is now connected to the Particle cloud, it does not know who it belongs to, so you need to "claim" it by logging into Particle.io and clicking the Devices icon (it looks like a target). Then click the ADD NEW CORE button and paste in the long number you copied earlier after issuing the I command (Figure 9-2). Finally, click CLAIM A CORE. You will then be prompted for a name for the Photon/Core.

You should now be able to use your Photon/Core as normal.

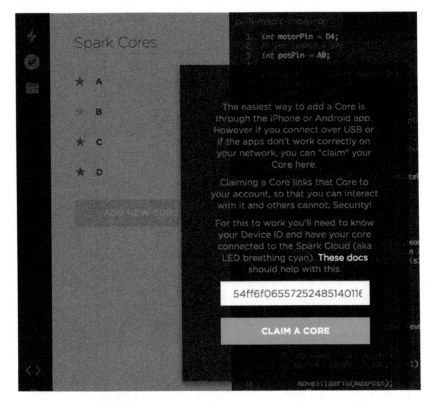

Figure 9-2. *Configuring a Photon/Core over USB*

Factory Reset

From time to time, your Photon/Core will misbehave and need resetting. A simple press of the Reset button or powering it on and off will restart the device and cure many problems; however, if the problem is something in your code, then it may stop you from flashing replacement code onto the app. This is especially common if you are experimenting with control of the WiFi or making the device sleep to save power.

When all else fails, you can perform a full factory reset. To do this, press and hold down both buttons on the device and then release the Reset button. There will be some yellow flashing and then eventually it will start to flash white. When it starts flashing white, you can release the Setup (Mode on the Core) button and wait while the device does a full factory reset.

Eventually, the white flashing will end and your device should be back to its original state. You will now have to tell it your WiFi credentials again.

Programming a Photon Using Particle Dev

I quite like the Web IDE and find it a convenient way to program and manage my devices. Particle also provides a desktop version of the IDE that you can download onto your computer. This is called Particle Dev and in many ways looks and operates in a very similar manner to the Web IDE.

At the time of writing, this tool is not fully offline, as it uses a web service to compile the app and flash it onto the device. However, this is a work in progress, and full offline programming of Photons and Cores over USB will be available in the future.

Debugging with the Serial Monitor

One problem with using an IoT device like the Photon or Core is that it can be tricky to work out what is happening when things go wrong with your code.

A common way to find out what is going on in a program is to add trace to your code. Trace is nothing more than program commands that print out some text somewhere so that you can see when the device runs a certain line of the program.

The problem with this on a device like the Photon or Core is that there is no screen on which the messages can appear. But what you can do is arrange for these messages to be sent over USB to a terminal application on your computer like Putty or the screen command that you used in "Configuring a Photon Using USB" on page 175.

As an example, try uploading the following app to your device. You can find the code in the file *ch_09_Trace_Example* in the *PHOTON_BOOK* library:

```
void setup() {
    Serial.begin(9600);
}
```

```
void loop() {
    if (Serial.available()) {
        char ch = Serial.read();
        if (ch == '?') {
            Serial.print("Hello millis()=");
            Serial.println(millis());
        }
    }
}
```

Now open screen or Putty just as you did in "Configuring a Photon Using USB" on page 175. Enter ? in the terminal window once it's connected, and you will see your device report the number of milliseconds since it last rebooted. Press ? again to repeat the process.

```
Hello millis()=101285
Hello millis()=107317
```

To start the communication, the line `Serial.println(9600)` starts communication at the baud rate (communication speed) of 9600.

The `loop` function checks for an incoming communication from screen or Putty and if the character received is a ?, first a message string and then the result of calling `millis()` are printed over the serial connection.

The difference between `print` and `println` is that `println` starts a new line at the end of the value it prints out, whereas `print` does not.

The Electron

A new Particle product is the Electron. At the time of writing, this is still at the stage of a Kickstarter project, but this device will function very like a Photon or Core. Instead of communicating using WiFi, however, the Electron communicates using a cellphone-style GSM modem. This means that the device can be truly mobile. For more news and updates on this interesting device, check back on the Particle.io website.

Power Management

WiFi uses quite a lot of power (generally up to 200mA). To put that into perspective, this means that if you were powering your Photon or Core from a pack of four AA batteries, you might expect it to get about 10 hours' use out of the batteries. With the WiFi turned off, you might easily get ten times as much time before your batteries are empty.

You can arrange for your device to put itself to sleep for a period of time to increase the battery life of the project by using the sleep command.

The sleep command takes a number of seconds as its parameter and turns WiFi off for that period of time. For example, to put WiFi to sleep for five seconds, you would use this:

```
Spark.sleep(5);
```

There are other commands to explicitly turn WiFi on and off, but remember that if WiFi is off, there is no way to flash a replacement app on the device, so you may well need to do a factory reset if things go wrong.

Check out the Particle documentation (*http://docs.particle.io/ firmware*) for the latest documentation on these features.

Summary

There are many more advanced features of the Photon and Core that will need to wait for a more advanced book. This book is, after all, a getting-started book. But, having mastered the basics, check out the Particle documentation, which is very thorough and well written. You will also find a helpful and informed community in the Particle forum (*http:// community.particle.io/*).

A/Parts

Component Suppliers

Photon and Core boards are available from a wide range of suppliers and from Particle itself. (Adafruit and Sparkfun Electronics are good suppliers for hobbyists.) You will also find kits,
shields, and other interesting Photon/Core-compatible modules.

Parts

Most of the parts used in this book are available in the Maker's
Kit available directly from Particle. This is by far the easiest way
to source your components. Please note that the contents of
the Maker's Kit may change, so check the contents list before
buying. All of the components are also available individually or
as parts of other kits from other suppliers. The following sections list these parts and will give you some help finding them.

The codes next to each part match up to the codes used in the
parts bins listed at the start of each project in the book.

Electronic Components

Even if you do not purchase the Particle Maker's Kit (referred to
as "Kit" in the remainder of this Appendix), the easiest way to
buy a basic set of electronic components is to buy a general
electronics starter kit. I have included Adafruit product codes
for some of the less common components.

Code	Description	Source
C1	220Ω resistor	Kit
C2	Red LED	Kit
C3	Tactile push switch	Kit
C4	Piezo buzzer	Adafruit: 160

C5	1kΩ resistor	Kit
C6	Photoresistor (1kΩ)	Adafruit: 161
C7	DS18B20 temperature sensor (encapsulated)	eBay, Adafruit: 381
C8	100nF (0.1uF) capacitor	Kit
C9	4.7k resistor	Kit. Included in Adafruit: 381
C10	Transistor 2N3904	Adafruit: 756
C11	1N4001 diode	Adafruit: 755

Modules and Shields

Code	Description	Source
M1	Relay shield	Particle.io
M2	PhoBot robot controller shield	monkmakes.com (available May 2015)
M3	HC-SR04 ultrasonic rangefinder	eBay
M4	Slide pot—motorized (10k linear taper)	SparkFun: 10976

Hardware and Connectors

Code	Description	Source
H1	DC jack-to-screw adapter (female)	Adafruit: 368
H2	DC jack-to-screw adapter (male)	Adafruit: 369
H3	Assorted multicore hookup wire	Adafruit: 1311
H4	Male-to-male jumper wires	Adafruit: 758
H5	Half-sized breadboard	Adafruit: 64
H6	Rover robot chassis kit (6V)	eBay. SparkFun: 12866
H7	Male 0.1 inch header pins	Adafruit: 392

Other

Code	Description	Source
Q1	12V 1A power supply with 2.1mm barrel jack	Adafruit: 798
Q2	12V DC electric bell	Hardware store

B/Photon LED Codes

The RGB LED of a Photon or Core uses different colors and rates of flashing to indicate what is going on inside the device. After a while, it almost feels like the Photon is demonstrating its current mood by what's going on in its LED.

Reset Sequence

When you reset a Photon, it will go through the following sequence on the LED:

1. Flashing blue: Listening mode, waiting for network information.
2. Solid blue: Smart Config complete, network information found.
3. Flashing green: Connecting to local WiFi network.
4. Flashing cyan: Connecting to Particle cloud.
5. High-speed flashing cyan: Particle cloud handshake.
6. Slow-breathing cyan: Successfully connected to Particle cloud.

Other Status Codes

- Flashing yellow: Bootloader mode, waiting for new code via USB or JTAG. (JTAG is a type of specialized programming hardware.)
- White pulse: Start-up, the Core was powered on or reset.
- Flashing white: Factory Reset initiated.
- Solid white: Factory Reset complete; rebooting.
- Flashing magenta: Updating firmware.
- Solid magenta: May have lost connection to the Particle cloud. Pressing the Reset (RST) button will attempt the update again.

Error Codes

Generally, red flashing indicates a problem:

- Two red flashes: Connection failure due to bad Internet connection.
- Three red flashes: The cloud is inaccessible, but the Internet connection is fine.
- Four red flashes: The cloud was reached, but the secure handshake failed.
- Flashing yellow/red: Bad credentials for the Particle cloud.

You can find out more about resolving these kinds of problems at Particle's support page (*http://support.particle.io*).

C/Photon and Core Pinouts

These diagrams (Figures C-1 and C-2) summarize the uses that each of the various pins can be put to on the Core and Photon devices.

Core

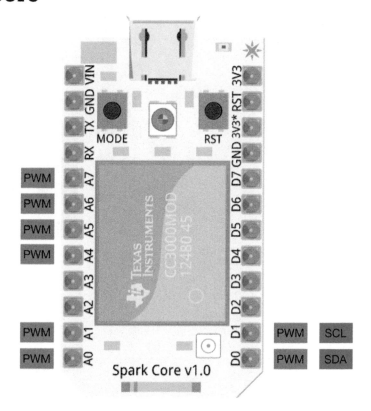

Figure C-1. *The Core pinout*

Photon

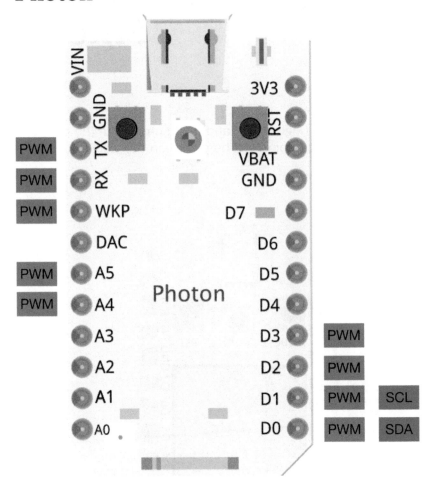

Figure C-2. *The Photon pinout*

About the Author

Dr. Simon Monk has a bachelor's degree in cybernetics and computer science and a PhD in software engineering. Simon spent several years as an academic before he returned to industry, co-founding the mobile software company Momote Ltd. He has been an active electronics hobbyist since his early teens. Simon is now a full-time author, and his books include *Programming Arduino*, *Programming the Raspberry Pi*, the *Raspberry Pi Cookbook*, *Hacking Electronics*, and various books in the TAB *Evil Genius* series. You can find out more about his books at *http://www.simonmonk.org* and follow him on Twitter (@simonmonk2).

Colophon

The cover and body font is Benton Sans, the heading font is Serifa, and the code font is Bitstream Vera Sans Mono.

CPSIA information can be obtained at www.ICGtesting.com
Printed in the USA
BVOW07s0328100615

403938BV00001B/1/P